IEE ELECTROMAGNETIC WAVES SERIES 8

SERIES EDITORS: PROFESSOR P. J. B. CLARRICOATS
G. MILLINGTON,
E. D. R. SHEARMAN
AND J. R. WAIT

Effects of the troposphere on radio communication

Previous volumes in this series

Effects of the troposphere on radio communication

M. P. M. HALL, M.Sc., C.Eng., M.I.E.E. M.Inst.P.
Rutherford and Appleton Laboratories
Science Research Council
Slough, England

PETER PEREGRINUS LTD.
on behalf of the
Institution of Electrical Engineers

Published by: The Institution of Electrical Engineers, London
and New York
Peter Peregrinus Ltd., Stevenage, UK, and New York

British Library Cataloguing in Publication Data

Hall, M. P. M.
Effects of the troposphere on radio communication.
— (Institution of Electrical Engineers. IEE
electromagnetic waves series; 8).
1. Tropospheric radio wave propagation
I. Title II. Series
621.3841'1   TK6553   79-42811

ISBN 0—906048—25—7

Composed at the Alden Press Oxford, London and Northampton
Printed in England by A. Wheaton & Co., Ltd., Exeter

Contents

Foreword

A knowledge of radiowave propagation is clearly essential to the development of all radio communication services. Although much assistance for the engineering and planning of such services can be obtained from a comprehensive body of empirical information based on extensive observations of propagation characteristics, even more can be gained through a thorough understanding of the scientific principles underlying the phenomena of propagation. In the field covered by this book the author has aimed to provide the basis for such an understanding.

Broadly speaking, and ignoring some deviations from the general pattern, the history of radio communication has involved the progressive exploitation of higher and higher frequencies, and along with this has gone a corresponding development of radiowave propagation studies both theoretical and experimental. These studies were concerned first with the effects of the surface of the earth on propagation; to these were soon added investigations of the ionosphere and its influence on propagation. Experimenters became increasingly aware of the influence of the troposphere on propagation (the main topic of this book) during the 1930s as research developed on transmissions in the VHF band; and this influence became very apparent during the Second World War and after as radio applications became widespread at even higher frequencies in the UHF, SHF and EHF bands; i.e. down to wavelengths as short as a few millimetres.

As the technology of transmission and reception has developed, together with advancing knowledge of wave propagation, more and more of the available radio spectrum has been exploited for all kinds of services, both terrestrial and space, and there are steadily increasing demands for frequency allocations. There is little sign that these demands will cease to expand even though full use may be made in the future, where appropriate, of waveguides and optical fibres in supplementation of lines and cables. When there are insufficient channels to meet requirements for radio communications in any part of the spectrum, then one must obviously resort to repeating frequency assignments if any attempt is to be made to satisfy the demand. The difficulty of doing this is now probably the greatest single problem facing radio communications planners, both national and international.

The ideal for the systems planner would be to achieve a service of 100% reliability with no susceptibility to interference from other services. When all other system parameters have been optimised it is propagation characteristics that play a dominant role in determining whether frequency sharing between (or within) services is possible. However, these characteristics are extremely variable with varying atmospheric conditions. In the event, therefore, the radio engineer must try to design a service for a close approach to 100% reliability with a near to zero susceptibility to interference. Economics cannot, of course, be excluded from the overall considerations, and prohibitive costs alone prevent the ultimate realisation of perfection, even if it were achievable on technical grounds. One therefore has to define limits of performance acceptability that may differ somewhat according to the kind of service concerned.

It is thus evident that careful planning of frequency allocation to services must be made on an international as well as national basis if optimum use is to be made of the radio frequency spectrum. International planning is mainly achieved through Administrative Radio Conferences of the International Telecommunication Union (ITU) on a worldwide or regional basis as appropriate, and at times on a particular service basis. The technical bases required for such administrative conferences stem largely from the studies carried out by the International Radio Consultative Committee (CCIR), which is one of the technical committees of the ITU. The CCIR pursues its work in a number of study groups that are concerned with various types of radio service together with other groups that are more basic in nature and deal with matters of interest to one or more service study groups; and among these basic groups are those concerned with radiowave propagation.

This book is concerned primarily with the effects of the troposphere and the ground on wave propagation at frequencies greater than about 30 MHz, although reference is made also to ionospheric phenomena where these are relevant. Tropospheric effects are directly related to weather and climatological conditions, and it is right that the author should devote considerable space to the many aspects of meteorology that enter into the study of tropospheric radiowave propagation. He is well acquainted with fundamental studies in this field through his research at the Appleton Laboratory. Furthermore, through his activities in CCIR Study Group 5 he is well aware of the practical problems of radio communication to which a knowledge of propagation must be applied, as will be evident from the many references in the text to CCIR studies. The reader should, therefore, find the book helpful in understanding the fundamental characteristics of radiowave propagation at frequencies in the VHF and higher bands and the consequence of taking account of these characteristics in the planning of radio communication at such frequencies.

J.A.Saxton, C.B.E., D.Sc., Ph.D., C.Eng., F.I.E.E., F.Inst.P.

Consultant, Home Office Directorate of Radio Technology, London. Formerly Director, Appleton Laboratory, Science Research Council, Slough. Chairman Study Group 5, International Radio Consultative Committee.

Preface

The last two decades have seen major changes in the demand for radio facilities and in the means for providing them. Perhaps the most important advances in recent years have been in satellite technology and the expansion into frequency bands above 10 GHz. Technological developments and an improved understanding of the limitations imposed by the atmosphere on radiowave propagation are leading to the development of systems capable of meeting the growth in communication requirements with a very high degree of reliability.

This book brings together in one volume an up-to-date account of tropospheric influences on radio communication and of certain terrain and ionospheric effects where appropriate. Following an introductory chapter, a description is given of the physical effects that influence radio services. The remaining and longer part of the book considers point-to-point links within and beyond the horizon and via space vehicles, as well as area coverage of a transmitter used for broadcasting or mobile application, and the various interference problems that may disrupt communications. Where these services are particularly affected by specific physical phenomena, cross reference to earlier chapters will be found. This fresh approach to the subject enables general principles to be considered first, avoids the duplication of basic information and leaves specific applications to be examined separately.

The book is intended to provide a general coverage of the field for those working on specialised aspects of radio communications, as well as those entering the field or engaged in formal studies of the subject. It is hoped that the form of the indexing, linking the contents with keywords used in the text and the extensive references to sources of material will make the work useful as a reference book.

It is difficult to select for acknowledgment only a few names from the many people to whom I am indebted. To have produced this book at all has needed the continuous support and assistance of my family, especially my wife. Also I must certainly thank here Dr. John A. Lane, Mr. George Millington and Dr. Harold T. Dougherty for their many constructive comments on the text, and many others at the Appleton Laboratory for their generous exchange of ideas over the years. The book has also benefitted from my long association with many colleagues both in the UK and in a number of other countries, whom it is a pleasure to have as friends.

My formal thanks too are no less sincere. In particular I am grateful to the Director of the Appleton Laboratory for permission to use the many Figures not specifically attributed to other sources, and to the Director of the CCIR for the use of the large number of Figures which are shown in the text as being taken from CCIR material. Figs. 4.15 and 6.5 are based on curves given by A. Picquenard in his book *Radio wave propagation,* published by Macmillan Press Ltd., and are reproduced with the publishers kind permission.

Martin P. M. Hall
Appleton Laboratory, July 1979

Chapter 1

Introduction

1.1 Propagation mechanisms through the troposphere

The *troposphere* is the lower part of the atmosphere, in which temperature generally decreases with height. It extends from ground level to an altitude of about 9 km at the earth's poles and 17 km at the equator. The height of the upper boundary also varies with atmospheric conditions: for instance, at middle latitudes it may reach about 13 km in anticyclones and decline to less than 7 km in depressions. It is in the troposphere that changes of temperature, pressure and humidity, as well as clouds and rain, influence the way in which radiowaves propagate from one point to another. Ionisation of atmospheric gases is negligible within the troposphere, but it is appreciable at heights of 60 to 1000 km, i.e. in the *ionosphere*. This region exerts a considerable influence on radiowaves at frequencies below about 30 MHz but seldom at higher frequencies. However, when this does occur it may cause interference between terrestrial radio terminals or fading on earth-to-space radio paths.

At frequencies above 30 MHz (*a*) localised refractive index fluctuations in the troposphere can *scatter* radio energy, (*b*) horizontally-stratified abrupt changes in refractive index can cause *reflection* and (*c*) extended negative gradients can cause *ducting*.

All these mechanisms can carry energy far beyond the normal horizon and so give rise to interference between one radio path and another. Reflection most affects frequencies between about 30 and 1000 MHz and ducting most affects frequencies above about 1000 MHz. Fortunately the latter occur very infrequently over land, although ducts often exist over sea. However, forward scattering of radio energy is sufficiently dependable that it may be used as a mechanism for long-distance communications, especially at frequencies of about 0·3 to

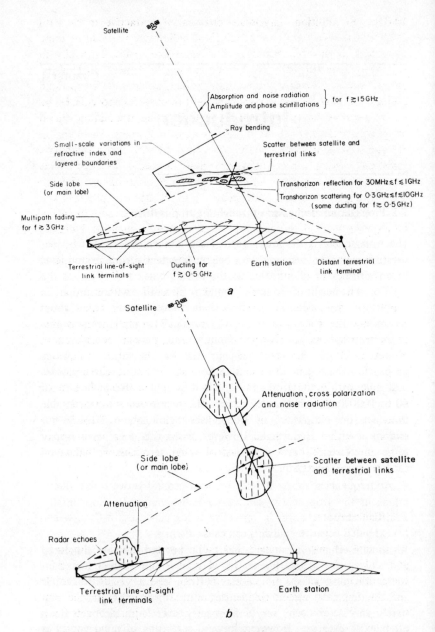

Fig. 1.1.   *Some effects of the troposphere on radiowave propagation*
(*a*) Effects of atmospheric gases (clear air) and associated refractive index changes
(*b*) Effects of cloud and precipitation (for $\gtrsim$ 3 GHz)

10 GHz. In addition, large-scale changes of refractive index with height cause *refraction* (ray bending) of radiowaves that can be quite significant at all frequencies at low elevation angles, especially in effectively extending the radio horizon distance beyond that of the optical horizon. Apart from these refractive-index effects, radio propagation may be strongly influenced at frequencies above 3 GHz by the presence of heavy *rain,* and above 15 GHz the attenuation caused by *oxygen and water vapour* in the air may be important, depending on the application. In addition, the absorption by rain and atmospheric gases will have an associated thermal noise emission. All these effects are indicated in Fig. 1.1*a* and *b*, but the effects of *terrain* are often no less important. At frequencies greater than 30 MHz, the presence and shape of hills has an important influence on the field strength of energy propagating beyond the horizon. At yet higher frequencies, buildings and other obstacles have a marked effect by diffraction, scatter and specular reflection mechanisms, when the wavelength is small compared with the dimensions of the obstacles.

The principal limitations and benefits to radio communications of various tropospheric propagation and terrain characteristics are shown in Column 2 of Table 1.1. Column 3 shows some of the *system considerations.* Both the propagation and system considerations influence the uses made of (or proposed for) the various frequency bands, and these are indicated in Column 4. The increased bandwidth and improved antenna gain available at the higher frequencies enable services to operate that could not make use of lower frequencies, e.g. those using high data rates. In addition, the increased bandwidth enables several or many signal information channels (e.g. telephone or television channels) to be combined in a single transmission that would require several lower frequency transmissions.

1.2 Radio services

In principle it may be argued that radiowave transmission should be used only if there is no alternative, where it is more economical or for some other special reason. Land, maritime and aeronautical mobile services, together with navigational aids, are obviously in the first category. With the wide use of car radio and hand-portable sets (even for television), broadcasting may also be considered first category. For fixed-terminal point-to-point trunk-routing of information, radio is appropriate where the distance between terminal points is large or the intervening terrain unaccommodating. The time is fast approaching

Table 1.1   Propagation characteristics, service allocations and system characteristics of the radio spectrum at VHF and greater frequencies

Wavelength frequency	Band*	Propagation characteristics	System characteristics	Service†
10 m 30 MHz	Metric VHF Band 8	Service area coverage by ground wave somewhat beyond horizon	Significant gain from simple (cheap) antennae. Several MHz bandwidth available per channel	Broadcasting (TV and high quality sound)
		Some protection by hills from distant unwanted transmissions	Moderately simple (cheap) receivers	Fixed service
				Land mobile
		Some interference via ionosphere		(Aeronautical mobile)
		Some interference via tropospheric layers		(Aeronautical navigation)
1 m 300 MHz	Decimetric UHF Band 9	Service area coverage less far beyond horizon	Higher gain antennae than at VHF for comparable size (and cost)	As above, plus: [Radar (radio location)]
		Protection by hills from distant transmissions better than VHF	More bandwidth per carrier	(Radio navigation)
		'Troposcatter' mode usable No interference via ionosphere	Moderately simple (cheap) receivers	
		Some interference via tropospheric layers and via ducting (at top of band)		

Table 1.1 *continued*

Wavelength frequency	Band*	Propagation characteristics	System characteristics	Service†
10 cm 3 GHz	Centimetric SHF Band 10	Diffraction losses limit systems to line-of-sight or 'troposcatter' links Screening and scatter by obstacles important Interference via ducting and precipitation scatter Rain attenuation becoming serious	Much higher gain paraboloidal antennae practical (but more complex) Hundreds of MHz bandwidth available per carrier	Satellite communications Radar (radio location) Fixed services; high-speed data or 'troposcatter' Land mobile (Radio navigation) (Broadcasting) (Satellite broadcasting)
1 cm 30 GHz	Millimetric EHF Band 11	Rain attenuation serious Atmospheric gaseous absorption becoming serious	Comparable high gain to SHF from smaller antennae Yet more bandwidth per carrier	As above, plus: (Intersatellite communications) (Aeronautical and maritime satellite services) (Local fixed services high data rate)

(continued)

Table 1.1 *continued*

Wavelength frequency	Band*	Propagation characteristics	System characteristics	Service†
1 mm 300 GHz		Rain attenuation serious	Yet smaller antennae for same high gain (but difficult to fabricate)	Not yet allocated above 275 GHz for satellite or 40 GHz for terrestrial
	Sub-millimetric Band 12	Strong gaseous absorption except for 'windows' of lower attenuation	Transmitters and receivers more difficult to fabricate	
0·1 mm 3000 GHz				

* These band numbers n, are used in the ITU Radio Regulations.[1] The frequency 10^n Hz is within the band
† All allocated bands except EHF include amateur, radio astronomy and space research services. Minor usage shown in parentheses

when demands on the radio spectrum will be so great that some form of cable transmission will have to provide for services with lower claims for using radio. In considering these radio services, brief mention must be included of frequencies less than 30 MHz for which long-distance communication can be achieved by means of ionospheric propagation.

Land-mobile services operate over distances from a few kilometres to worldwide ranges. Long-range requirements may be catered for by satellite (particularly at UHF) or by conventional HF radio links. Short-range requirements (often within line-of-sight) are best served with allocations in the VHF and lower UHF bands, where diffraction round local obstacles and slightly beyond the radio horizon is effective, and where fairly small antennae may be used efficiently.

For *maritime-mobile services,* MF and HF bands are allocated for telegraphy, telephony and teleprinter use, but the services provided are far from satisfactory owing to ionospheric propagation problems, interference from other services and the inadequacy of the present allocation to meet the traffic demand. Three small bands at VHF carry a multiplicity of services, such as public correspondence, ship-to-ship calls, port operations, ship movement etc., and six frequencies in the UHF band have been allocated for use 'on board'. Two bands are available at 1·6 GHz for communication via satellites, initially for telephone and teleprinter services, and development of the low interference and wideband opportunities of these satellite services are likely to transform maritime communications and navigational aids over the next one or two decades.

Aeronautical-mobile allocations are scattered throughout the HF band, and at present provide the only means of communicating directly with aircraft over long distances. Here again, these services suffer from propagation problems and interference from other users. VHF and UHF links provide the main ground-to-air line-of-sight communications to aircraft. The use of satellites for communication and navigational aids to aircraft at 1·6 GHz is expected to provide greatly improved services.

Examples of *radionavigation and radiolocation systems* can be found throughout the frequency spectrum. Navigational aids use

(*a*) VLF, LF and MF bands for wide-coverage position-fixing systems
(*b*) the HF and VHF bands for a large number of maritime and aeronautical beacons, radio-direction finding, distance measuring and guidance systems and
(*c*) microwave frequency bands for the majority of the more sophisticated equipment and the proposed satellite-based wide-area navigation services. The majority of radars operate between 200 MHz and 35 GHz.

Broadcasting accounts for nearly 60% of the spectrum up to 1 GHz (i.e. when examined decade by decade). The LF band is suitable for long-distance ground-wave propagation, but only in those parts of the world where atmospheric noise is moderate to low. In the MF bands there are many thousands of broadcasting transmitters, but, due to mutual interference between these, the coverage at night hardly exceeds that of a VHF station. Broadcasting in the lower HF band is more attractive than the MF band in the tropics because good coverage of a large area may be achieved with relatively low atmospheric noise. The upper end of the HF band is used for long-distance external broadcasting. The VHF/UHF broadcasting bands, with slight differences in various parts of the world, are allocated as follows:

Band I	41–68 MHz
Band II	87·5–100 MHz
Band III	174–216 MHz
Band IV	470–582 MHz
Band V	614–960 MHz

Band I, television, has good area coverage compared with UHF, but provides only five 5 MHz-wide channels. Furthermore, interference from distant stations presents a severe problem during summer months because of sporadic E-layer propagation. Sound broadcasting in Band II has advantages over lower frequencies in that the wider bandwidth accommodates frequency modulation (giving a better signal/noise ratio and greater dynamic range) as well as stereophonic and quadraphonic broadcasting. For Band III television, and more so at UHF in Bands IV and V, it is possible to have a smaller directional antenna, but diffraction losses produce more shadowed areas that have to be filled with large numbers of relay stations. By international agreement, Bands IV and V accommodate 44 channels, each 8 MHz wide, numbered to a common system. Plans are well advanced in many countries for the introduction of broadcasting stations in the 11·7–12·5 GHz band, both by satellite and terrestrially.

Until a few years ago, the HF band provided almost all intercontinental *fixed-terminal point-to-point* services and a modest communications capacity for many shorter links. They are still fairly heavily used in regions where the more reliable submarine cables, satellite and radio-relay systems are not available. In developed countries substantial use is made of frequencies of about 1 or 2 GHz for tropospheric-scatter radio systems, consisting of links of a few hundred kilometres. These systems are used particularly for the transmission of an information bandwidth of a few hundred kilohertz with high reliability. By contrast, line-of-sight microwave radio-relay systems carry a large part of the main trunk public telephone and video circuit networks, some 1800 telephone channels (each of 4 kHz) occupying the bandwidth of one TV channel. They also carry many high-capacity private digital-data networks. In many countries such systems extensively occupy bands between 2 and 10 GHz, and broad bands even higher in the frequency spectrum are now being taken into use. Satellite systems for international communication already occupy the 4/6 GHz (downlink/uplink) frequency bands that are shared with terrestrial point-to-point stations. Since these bands are becoming saturated in the busier parts of the geostationary satellite orbit, the 11/14 GHz bands are being developed, and extension of satellite services into the 20/30 GHz bands is envisaged in the near future. The predominant use of these systems is for long-distance telephone communications, but applications such as flexible or multiterminal wideband services (e.g. data networks or television distribution) are also emerging.

Because of the ever increasing demand for radio facilities, the *allocation of the radio spectrum* between radio services must be planned carefully and reviewed periodically. Some mention is made in the Foreword of how this is carried out at World Administrative Radio Conferences convened by the International Telecommunications Union. Most of these conferences relate to specific problems, e.g. space technology (1971), maritime mobile (1974), satellite broadcasting (1977), aeronautical mobile (1978), but occasionally (e.g. 1979) they are charged with a revision of the whole international frequency allocation table and many other aspects of international regulation of the use of the radio spectrum. In addition, certain Regional Conferences examine specific regional issues. The ITU has some 150 member countries and they are collectively responsible for formulating radio regulations[1] on the manner in which radio systems and services shall be operated, the use of frequencies for different services and a wide range of technical criteria that should be observed. Within the ITU, the International Radio Consultative Committee (CCIR) produces reports designed to provide an up-to-date account of the state of knowledge in various fields relating to radio communications and recommendations for procedures to be adopted in pursuing specific technical problems. The content of some of these authoritative documents is pertinent to the subject material of this book and many references are made to them in what follows.

1.3 Quantifying propagation performance

In order to plan reliable radio systems it is important to be able to predict system-loss/time-probability statistics on radio paths. This enables complete systems to be designed without over or under specifying the transmitter and receiver characteristics or those of their antennae. For similar reasons it is necessary to be able to predict such statistics for radio paths to and from other terminals where co-channel interference may occur.

When considering quantitatively the attenuation on radio paths through the troposphere under varying conditions, it is necessary to adopt a simple *standard of reference*. The customary standard is a theoretically calculated loss for waves propagated in free space between two hypothetical antennae of known characteristics. Isotropic antennae are usually adopted because they are most fundamental in that they radiate, or receive, radio energy uniformly in, or from, all directions. The 'directive gain' of an actual antenna can then be judged against the theoretical ideal. This gain is the ratio of the power flux per unit area at a distance from the antenna transmitting a given power to the power flux per unit area that would exist at the same distance from an isotropic antenna transmitting the same power. The commonly stated gain G of an antenna is the 'directivity gain', i.e. the maximum directive gain, but a knowledge of how this gain decreases as the antenna is moved off the optimum orientation (i.e. the radiation pattern) is required for calculating off-axis co-channel interference or multipath effects.

The standard (theoretical) reference for calculating radio propagation losses over radio paths in the atmosphere is the *basic free-space transmission loss L_{bf}*, against which actual transmission results can be expressed as a ratio. It is the ratio of the radio frequency power radiated in free space from an ideal loss-free isotropic ($G = 1$) transmitting antenna to the power available from an ideal loss-free isotropic receiving antenna.[301] Since the power flux per unit area P_a(W/m^2) at a distance $d(m)$ from a transmitter radiating a power $P_t(W)$ from an isotropic loss-free antenna is given by

$$P_a = P_t/4\pi d^2 \tag{1.1}$$

and the power available from a loss-free antenna P_r is the product of P_a and the effective aperture A_e, where this area is related to the antenna gain by

$$A_e = G\lambda^2/4\pi \tag{1.2}$$

then the basic free-space transmission loss follows from eqns. 1.1 and 1.2, i.e.:

$$L_{bf} = P_t/P_r = (4\pi d/\lambda)^2 \tag{1.3}$$

where d and λ are in the same units. This equation is sometimes expressed in logarithmic terms as

$$L_{bf} = 32{\cdot}4 + 20 \log_{10}d + 20 \log_{10}f \quad \text{dB} \tag{1.4}$$

Note that doubling the distance d(km) or the frequency f(MHz) increases the loss by 6 dB. Also, for a parabolic dish antenna, A_e is less than $\pi D^2/4$ since the illumination or collection by the antenna feed is not completely uniform across the diameter D and the antenna shape cannot be a perfect parabola. The ratio of the effective aperture and physical aperture is the '*efficiency*' of the antenna, and by the same factor the gain is less than ideal.

For a radio link in the atmosphere, the *transmission loss L* of a radio link is defined as the ratio of the radio frequency power radiated from the transmitting antenna to the resultant radio frequency signal power that would be available from the receiving antenna if there were no circuit losses other than those associated with radiation resistance.[301] The *basic transmission loss L_b*, sometimes called the *path loss*, is defined as the transmission loss expected between ideal, loss-free isotropic transmitting and receiving antennae.[301] The terms L_b and L (when expressed in decibels) are related by

$$L = L_b - (G_t + G_r - L_d) = L_b - G_p \quad \text{dB} \tag{1.5}$$

where G_t and G_r are the free-space antenna gains (with respect to isotropic) and the *path antenna gain G_p* is less than the sum of the plane-wave gains by the *aperture-to-medium coupling loss* (or gain degradation) of the antennae L_d. This loss occurs when the operating environment of the two antennae causes the wavefront at the receiving antenna to be other than plane (see Section 6.4). Where the transmitting and receiving antennae themselves have circuit losses L_t and L_r (excluding transmission line losses), it is sometimes convenient to consider a *system loss L_s* such that

$$L_s = L + L_t + L_r \quad \text{dB} \tag{1.6}$$

However, for the types of antennae used at frequencies above 30 MHz it is usually preferable to include these losses when establishing the measured values of G_t and G_r so that they become part of the efficiency factor mentioned above.

The *path attenuation,* that is the loss relative to the free space value, is given by

$$A = L_b - L_{bf} = L - L_{bf} + G_t + G_r - L_D \quad \text{dB} \quad (1.7)$$

where L, G_t and G_r may be measured and L_{bf} may be calculated. In subsequent chapters there will be considerable mention of the 'specific attenuation' γ(dB/km), which should be distinguished from the term 'attenuation coefficient' α (Nepers/km) used in some other texts. The path attenuation is the sum of losses due to scintillation, multipath effects, rain and atmospheric gases, the relative importance of which, and indeed whether they are relevant at all, depends on the frequency, the path geometry and the prevailing weather conditions.

In certain applications, notably when considering the potential service area coverage of a transmitter for broadcasting or mobile radio, it is convenient to express measurements in terms of *field strength* rather than transmission loss. Clearly the two are closely related. The RMS field strength E(V/m) at a point where the power density of a plane wave is P_a (W/m^2) is given by

$$E = (120\pi P_a)^{1/2} \quad (1.8)$$

In terms of the power P_r(N) available from a loss-free isotropic antenna, then, from eqns. 1.1, 1.3 and 1.8

$$E = (480\pi^2 P_r/\lambda^2)^{1/2} \quad (1.9)$$

Also, from eqns. 1.1, 1.3 and 1.8, the field produced by a transmitter of equivalent isotropically radiated power (EIRP) P_t(W) at a distance d(m) in free space (with d sufficiently large for the wavefront to be considered plane) is given by

$$E_f = (30P_t/d^2)^{1/2} \quad (1.10)$$

An alternative formula for the free-space field, used when d is expressed in kilometres, P_t in decibels > 1 kW, and with a transmitting antenna gain G_t in decibels, is

$$E_f = 105 + P_t + G_t - 20 \log_{10} d \quad \text{dB} > 1 \,\mu\text{V/m}$$
$$(1.11)$$

Using these units, the field strength E dB $> 1 \,\mu$V/m for non-free-space conditions is related to the basic transmission loss by

$$L_b = 137 + 20 \log_{10} f + P_t + G_t - E \quad \text{dB} \quad (1.12)$$

Often P_t is quoted in units of decibels > 1 kW EIRP, in which case $G_t = 0$ dB, but when calculating field strength (or transmission loss) for specific radio services, it is sometimes preferable to take as a reference an idealised antenna which is closer to that actually used than is an isotropic antenna.

Two such reference antennae are a *halfwave dipole* (with a gain of $2 \cdot 2$ dB with respect to an isotropic antenna), and a *short vertical unipole above a perfectly conducting ground plane* (having a gain of $4 \cdot 7$ dB with respect to an isotropic antenna for uniform current in its length less than $\lambda/8$). Like the monopole, a vertically-oriented halfwave dipole will radiate omnidirectionally in a horizontal plane with vertical polarisation (electric vector).

In all real situations the antennae will be of a more complex nature. Operation from a fixed site will normally enable high-gain antennae to be used to obtain maximum wanted signals from a known direction, and a minimum of any unwanted signals that may be on the same frequency in another direction. By way of examples, a multi-element 'yagi' antenna as used for VHF television reception will normally have a gain of less than 15 dB with respect to an isotropic antenna, whereas one of the large paraboloidal SHF antennae normally used at an earth station on an earth-to-satellite link would have a gain of about 50 dB with respect to an isotropic antenna.

Many aspects of successful propagation of radiowaves through the troposphere (at VHF and higher frequencies) are largely dependent on the weather, which in most climates is very variable. In practice, an essentially *statistical approach* is adopted for the prediction of propagation effects. Although data collected on one radio path during one time period may be very different from data collected over an apparently similar path over a similar time interval, data collected over periods of years do give statistical generalisations that enable radio engineers to design services with a reliability factor close to 100% and a near to zero vulnerability to interference. For this to be achieved, it is necessary to apply statistical data with an understanding of how they were derived and an appreciation of the influences of various meteorological and terrain conditions on radiowaves. These influences are examined in the following chapters in the light of current knowledge.

Chapter 2

Atmospheric refractive index

2.1 Introduction

Changes in refractive index n of only a few parts in a million can have an important effect on radiowaves and as values are so near to unity (typically $1\cdot00035$) it is usually more convenient to work in parts per million above unity N, i.e.

$$N = (n - 1) \times 10^6 \qquad (2.1)$$

which is strictly the refractivity, but more usually referred to as refractive index.

The percentage composition of clean dry air, by volume, for the principal gases is nitrogen 78·1, oxygen 20·9, argon and other rare gases 0·9 and carbon dioxide 0·03. This composition is practically constant up to 50 km or more, i.e. well above the top of the stratosphere. By contrast, the quantity of water vapour present is highly variable. This is important since the permanent dipole moment of the molecules of water vapour cause it to be a very significant contributor to the variability of the atmospheric refractive index. Typically the proportion, by volume, of water vapour in the air at ground level ranges from less than 0·001% in the arctic to more than 5% in the tropics. However, this proportion decreases rapidly with height and is highly dependent on local air temperature.

A generally accepted relationship for the dependence of atmospheric refractive index on the pressure P(mb), the temperature T(K) and the water vapour pressure e(mb) is[315]

$$N = 77\cdot6P/T + 3\cdot73 \times 10^5 e/T^2 \qquad (2.2)$$

This equation is correct to within 0·5% for atmospheric pressures between 200 and 1100 mb, air temperatures between 240 and 310 K, water vapour pressures less than 30 mb and for radio frequencies less than 30 GHz. However, outside the absorption bands discussed in Section 3.4 it is generally useable for frequencies up to 1000 GHz. The equation is semi-empirical, its form being derived from consideration of the effects of polar and non-polar gases and the constants being obtained from measurements.[2] The ideal gas law is assumed and dispersion is neglected.

The terms $77 \cdot 6 P/T$ and $3 \cdot 73 \times 10^5 e/T^2$ are sometimes referred to as N_{dry} and N_{wet}, respectively, and some indication of the magnitude of these two contributions to N, and their variation with temperature and relative humidity, is given in Table 2.1 (for an atmospheric pressure of 1000 mb). At very low temperatures N_{wet} becomes very small even for saturated air, and so N is almost independent of relative humidity. As the temperature rises, there is a slow decrease in N_{dry} but a rapid increase in the saturated value $N_{wet\,max}$. At high temperatures $N_{wet\,max}$ can become somewhat larger than N_{dry} and so N varies considerably with relative humidity. At high temperatures and high relative humidity, N is very sensitive to small changes in temperature and relative humidity. Consequently, the variability of refractive index in tropical areas is far greater than that of cold climates.

Table 2.1 also includes a column of values of e_s, the maximum possible (saturated) vapour pressure at the air temperature t° C in question, where

$$e_s = 6 \cdot 11 \exp \left[19 \cdot 7 t/(t - 273) \right] \qquad (2.3)$$

since for any relative humidity $H\%$

$$e = He_s \qquad (2.4)$$

(The form of eqn. 2.3 may also be used to relate e and the dew point t_D in place of e_s and t.)

It is sometimes preferable to consider the mass of water vapour per unit volume $m(g/m^3)$ in the air, where this mass is the parameter most closely related to the effect under study (see, for example, the end of Section 3.4). This mass is normally referred to as the *water vapour concentration*, and sometimes as the vapour density or absolute humidity. It is related to water vapour pressure and temperature by

$$m = 216 \cdot 7 e/T \qquad (2.5)$$

The probability of a certain water vapour concentration occurring is dependent on the air temperature, and Fig. 2.1 shows this dependence based on data collected at ground level four times a day over 10 years at a typical site in the UK.[2] The Figure shows measured values of m

Table 2.1   Variation of refractive index N with temperature $t°C$ and relative humidity $H\%$ for an atmospheric pressure of $1000\,mb$ *

$t°C$	H%						Maximum value of N_{wet}	e_s
	0	20	40	60	80	100		
	N_{dry}							mb
−30	319·3	320·0	320·7	321·4	322·1	322·7	3·4	0·5
−28	316·7	317·5	318·3	319·1	319·9	320·7	4·0	0·6
−26	314·2	315·1	316·1	317·0	317·9	318·9	4·7	0·8
−24	311·6	312·8	313·9	315·0	316·1	317·2	5·5	0·9
−22	309·2	310·5	311·7	313·0	314·3	315·6	6·5	1·1
−20	306·7	308·2	309·7	311·2	312·7	314·2	7·5	1·3
−18	304·3	306·1	307·8	309·6	311·3	313·1	8·8	1·5
−16	301·9	304·0	306·0	308·0	310·1	312·1	10·2	1·8
−14	299·6	302·0	304·3	306·7	309·0	311·4	11·7	2·1
−12	297·3	300·0	302·7	305·5	308·2	310·9	13·6	2·5
−10	295·1	298·2	301·3	304·4	307·5	310·7	15·6	2·9
−8	292·8	296·4	300·0	303·6	307·2	310·8	17·9	3·4
−6	290·6	294·8	298·9	303·0	307·1	311·2	20·6	3·9
−4	288·5	293·2	297·9	302·6	307·3	312·0	23·5	4·6
−2	286·3	291·7	297·1	302·5	307·8	313·2	26·9	5·3
0	284·2	290·4	296·5	302·6	308·7	314·9	30·6	6·1
2	282·2	289·1	296·1	303·1	310·0	317·0	34·8	7·0
4	280·1	288·0	295·9	303·8	311·7	319·6	39·5	8·1
6	278·1	287·1	296·0	305·0	313·9	322·9	44·7	9·3
8	276·2	286·3	296·4	306·5	316·6	326·7	50·6	10·7
10	274·2	285·6	297·0	308·4	319·9	331·3	57·1	12·2
12	272·3	285·1	298·0	310·9	323·7	336·6	64·3	14·0
14	270·4	284·8	299·3	313·8	328·2	342·7	72·3	15·9
16	268·5	284·7	301·0	317·2	333·4	349·7	81·1	18·1
18	266·7	284·9	303·0	321·2	339·4	357·6	90·9	20·6
20	264·8	285·2	305·5	325·9	346·2	366·6	101·7	23·4
22	263·1	285·8	308·5	331·2	354·0	376·7	113·6	26·5
24	261·3	286·6	312·0	337·3	362·7	388·0	126·7	29·9
26	259·5	287·8	316·0	344·2	372·4	400·6	141·1	33·8
28	257·8	289·2	320·6	351·9	383·3	414·7	156·9	38·1

Table 2.1 *continued*

t° C	H%						Maximum value of N_{wet}	e_s
	0	20	40	60	80	100		
	N_{dry}							mb
30	256·1	290·9	325·8	360·6	395·4	430·3	174·2	42·8
32	254·4	293·0	331·7	370·3	408·9	447·5	193·1	48·1
34	252·8	295·5	338·3	381·0	423·7	466·5	213·7	53·9
36	251·1	298·4	345·6	392·9	440·1	487·4	236·2	60·4
38	249·5	301·7	353·8	406·0	458·1	510·3	260·8	67·5
40	247·9	305·4	362·9	420·4	477·9	535·4	287·5	75·4

[*] For a given temperature N_{dry} is proportional to pressure and N_{wet} is proportional to relative humidity
N_{dry} (at 0% rh) and the saturated (maximum) value of N, N_{wet} (at 100% rh), are indicated at each temperature, as are values of the saturated water vapour pressure e_s (mb)

exceeded for certain time percentages for each 2° C temperature interval. For the following discussion it also shows the relationship between water vapour concentration, relative humdity and temperature (from eqns. 2.3, 2.4 and 2.5). Evidently the long-term median value of m, $7·7 \text{ g/m}^3$, cannot occur below 7° C (except under super saturated conditions) but may occur for 50% of time above 9° C. Similarly, the value exceeded for only 0·1% of time in the long-term cannot occur below 19° C but is present for more than 1% of time above 21° C. For an air temperature below about 15° C, the air is almost completely saturated for 1% of time and is above 80% humidity for almost 50% of time. In the climate considered the air does not take up increasing water vapour concentrations at temperatures above about 20° C, and so the relative humidity values drop considerably, and the highest values of refractive index shown in Table 2.1 do not occur. The amount of water vapour occurring at ground level appears in certain atmospheric models used in later chapters. Normally, a median value of $7·5 \text{ g/m}^3$ is used, but it may be useful sometimes to refer to worldwide maps of the surface value that have been prepared for the different seasons of the year.[2, 4, 316]

It is convenient here to mention briefly two other parameters sometimes used by the radio meteorologist, namely the humidity mixing ratio and the specific humidity. The *humidity mixing ratio r* is the ratio of the mass m_v of water vapour in a volume to the mass m_a of dry air in the same volume, i.e.

$$r = m_v/m_a = 0.622e/(P-e) \qquad (2.6)$$

where 0·622 is the ratio of the densities of water vapour and dry air. It is usually expressed in the form $r = 622e/(P-e)$g/kg.

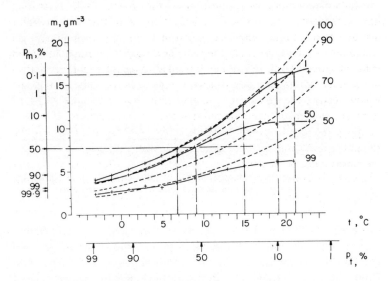

Fig. 2.1. *Probability of occurrence of water vapour concentration m as a function of temperature t at an inland site in the UK*
——— measured *m* values exceeded for 99, 50 and 1% of time
- - - - - - - *m* as a function of *t* for 100, 90, 70 and 50% relative humidity
P_m measured percentage of time that *m* is exceeded (all temperatures)
P_t measured percentage of time that *t* is exceeded (all vapour concentrations)
– – – points referred to in the text

The *specific humidity*, moisture content, or mass concentration q is the ratio of the mass of water vapour in a volume to the mass of moist air in the same volume, i.e.

$$q = m_v/(m_v + m_a) = r/(r+1) \qquad (2.7)$$

2.2 Measuring refractive index

The refractive index of the air, a characteristic which is so fundamental to many aspects of radio propagation, may be measured using many techniques, and to some extent those selected will depend on the purpose. Accuracy may be important in a laboratory, but in radiosondes used for routine upper air measurements, ruggedness, repeatability without recalibration and a wide range of response of the sensors may be more important. A degree of compromise is usually required for radiometeorology since most often the interest is in measuring how refractivity changes with height, using a relatively lightweight instrument suspended below a tethered or free-release balloon, and the techniques used are generally the same as those used for measurements made at ground level. If a system is needed to measure rapid refractive-index fluctuations, at a rate of several tens of hertz, only instruments which record the refractive index directly have adequate speed of response; such instruments are referred to as 'refractometers'. If a lower sampling rate can be permitted, or a lightweight system is essential, the refractive index may be determined indirectly from calculations based on measurements of total pressure, temperature and water vapour (partial) pressure of the air, using eqn. 2.2. The water vapour pressure is obtained by measuring the relative humidity, the dew point or the wet-bulb depression, and there has been much progress in means of measurements of these parameters.[5]

Microwave refractometers have been the most commonly used instruments for measuring refractive index directly. The principle of operation is to measure the change δf in resonant frequency f of a cylindrical cavity with partly open ends, which occurs due to a change δn in refractive index of the air passing through the cavity. Then, for a fixed cavity size

$$\delta n = -\delta f/f \tag{2.8}$$

Two types of microwave refractometers were first used in 1950.[6, 7] Both employed a 10 GHz resonant cavity with a substantial part of the end walls removed, so as to enable air to pass through freely. Such a cavity is shown in Fig. 2.2. The means of measuring the resonance frequency was different in the two types, but each has since been used extensively in aircraft, on tall masts and suspended from tethered balloons. More recently the use of solid-state circuitry and solid-state microwave sources has reduced the weight very considerably and reduced the short-term errors to (typically) less than 0·01 N-unit for refractive index fluctuation frequencies up to 30 Hz. A direct digital readout of refractivity may be obtained with a stability of 1 part in 10^7 per month[8, 9] and it is possible to use a similar instrument with smaller sampling cavities operating at about 20 GHz.

Non-microwave refractometers have been produced to operate at 10 MHz and at 403 MHz. In the former group of instruments[10] the sensing device was an air capacitor which formed part of a reasonant circuit in a high frequency oscillator, the total weight of the sondes being only 3·2 kg. In the latter,[11] open-ended metalised ceramic cavities were used and the sondes were dropped from aircraft.

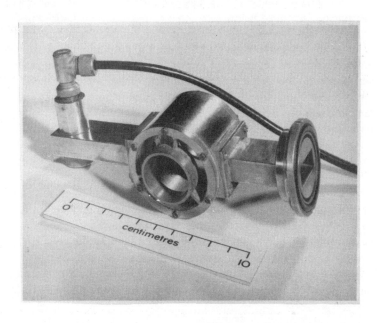

Fig. 2.2   *Microwave refractometer sampling cavity*

Before considering the many means that have been developed to *measure pressure, temperature and humidity simultaneously* in order to compute the refractive index, it is worth examining the accuracy required in terms of the differentials of eqn. 2.2. For typical atmospheric conditions near ground, i.e. $P = 1000$ mb, $T = 288$ K, $e = 11·9$ mb, (and so $H = 70\%$)

$$\delta N = 0·27\delta P - 1·3\delta T + 4·5\delta e \qquad (2.9)$$

Normally the variability of e has most effect on N, but sometimes variation in T can also have a considerable effect. The water vapour pressure e and its variations may be measured in terms of the dewpoint temperature T_D K, the wet-bulb temperature depression $D_w = T - T_w$ K or the relative humidity $H\%$. The wet-bulb depression is related to water vapour pressure by

$$e = e_{sw} - \beta P D_w \tag{2.10}$$

where e_{sw} is the saturated vapour pressure at the wet-bulb temperature and β is a term dependent on the wet-bulb temperature and its rate of aspiration as

$$\beta = A(0 \cdot 00115 T_w + 0 \cdot 686) \times 10^{-4} \tag{2.11}$$

where for $T_w \geqslant 273 \text{ K}$ A is 6·60 for a fully aspirated wet bulb and 7·93 for a naturally aspirated wet bulb (e.g. in a Stephenson screen). The accuracy required for different techniques in making indirect measurements of refractive index may be deduced from the differentials of eqns. 2.3, 2.4 and 2.10 and from eqn. 2.9 as

$$\delta N = 0 \cdot 27 \delta P - 1 \cdot 3 \delta T + 3 \cdot 6 \delta T_D \tag{2.12}$$

$$\delta N = 0 \cdot 27 \delta P - 4 \cdot 4 \delta T + 7 \cdot 2 \delta T_w \tag{2.13}$$

$$\delta N = 0 \cdot 27 \delta P + 2 \cdot 8 \delta T - 7 \cdot 2 \delta D \tag{2.14}$$

$$\delta N = 0 \cdot 27 \delta P + 1 \cdot 2 \delta T + 0 \cdot 77 \delta H \tag{2.15}$$

Conventional radiosondes, as used for weather forecasting, have been developed as moderately robust, low cost, throw-away devices to measure pressure, temperature and humidity at heights up to some 40 km; but the sensitivity and rapidity of response are often inadequate for radiometeorological studies for which detailed information in only the first few kilometres above ground is usually required. There has, therefore, been considerable development of *specialised radiosondes* which have been used suspended below tethered or free balloons for specific research applications. Many of these sondes have employed direct measurement of dewpoint, using mirror surfaces cooled either by the Peltier effect or by spraying on a coolant liquid. By these means temperature errors as low as 0·3 K and time constants as low as 1·3 s have been achieved. There has also been a great improvement in means of measurement of humidity using sensors which change electrical resistance according to the amount of water absorbed on their surfaces. The carbon hygristor has proved the most successful to date, and this technique could lead to a high-accuracy commercial sonde. Many of the specialised radiosondes have used the wet-and-dry bulb (psychrometer) principle. Psychrometric radiosondes have the advantage of being relatively simple, inexpensive, lightweight and easily calibrated, but they cannot be used much below the freezing point. An important point to remember about almost all radiosonde measurements of refractive index change with height is that the time constant (lag) of the measuring instrument will cause an underestimate of the rate of change of refractive index gradient. This is important when the radiosonde passes rapidly through layers of markedly different refractive index. The height h(m) of the radiosonde may be determined from the pressure reading by using the relationship

$$P_h = P_0 \exp\left(-gh/R\bar{T}\right) \tag{2.16}$$

where P_0 is the pressure at $h = 0$, $\bar{T} = \frac{1}{2}(T_h + T_0)$, the gravitational constant $g = 9{\cdot}81\,\text{m/s}^2$ and the specific gas constant $R = 287{\cdot}0\,\text{m}^2\,\text{s}^{-2}\,\text{K}^{-1}$. Typically at ground level the rate of decrease of atmospheric pressure with height is $-P_h g/RT = -119\,\text{mb/km}$ and in the first kilometre the decrease in pressure is 113 mb.

2.3 Variability of refractive index with height

One of the most significant parameters in the influence of the troposphere on radiowave propagation is the large-scale variation of refractive index with height, and the extent to which this changes with time. Variation of refractive index in the horizontal is negligible by comparison. Such localised changes of refractive index with height as cause ducting and reflection from elevated layers will be considered in Sections 2.5 and 2.6, and the small-scale variations that cause scatter will be considered in Section 2.7.

In practice, the measured *median of the mean gradient* in the first kilometre above ground in most temperate regions is about $-40\,\text{N/km}$, the pressure decrease with height accounting for about 75% of the total, and the normal humidity decrease and temperature decrease producing about $-17\,\text{N/km}$ and $+8\,\text{N/km}$, respectively. Some indication of the extent to which the statistical distribution of this mean gradient varies when measured over different height intervals is indicated in Fig. 2.3. This is based on a study carried out in the UK using data from 1273 soundings made with a specialised form of radiosonde over a period of three years.[12] The height intervals were chosen to meet various general path requirements, e.g. data for the 0 to 75 m interval may be most useful for short paths with low terminal heights, whereas the 0 to 1000 m interval may be appropriate for some paths extending beyond the horizon. Clearly, as the height interval becomes larger, the range of gradient values becomes less, but in each case the median gradient is $-40\,\text{N/km}$. Similar data have been collected on a worldwide basis.[13]

The decrease of refractive index with height h, through the whole troposphere, may be approximated by

$$N(h) = N_s \exp(-h/h_0) \tag{2.17}$$

where N_s is the surface value of refractive index and h_0 is a 'scale height'. The *mean refractive index gradient in the first kilometre height* ΔN, a much-used radiometeorological parameter, is then given by

$$\Delta N = N(1) - N_s = -N_s\{1 - \exp(-1/h_0)\} \tag{2.18}$$

where ΔN is negative for most time percentages.

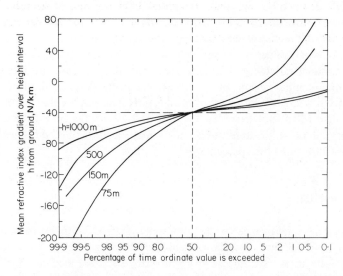

Fig. 2.3. *Distribution of mean refractive index gradient with height interval above ground level*

In temperate climates the monthly median values of N_s and ΔN vary from about 300 to 350, and -35 to -55, respectively, according to location and season. The CCIR define an average atmosphere as one in which $N_s = 315$ and $\Delta N = -40$, i.e.

$$N(h) = 315 \exp(-h/7 \cdot 36) \tag{2.19}$$

Several studies have shown ΔN to be inversely correlated with N_s,[14–17] and the general form is

$$\Delta N = -A \exp(BN_s) \tag{2.20}$$

where $2 \cdot 1 < A < 9 \cdot 3$ and $0 \cdot 0045 < B < 0 \cdot 0094$ according to climate.

World maps have been prepared giving the mean values and seasonal variations of refractive index reduced to sea level N_o, and the mean gradient in the first kilometre ΔN.[2, 18] On a worldwide scale it is clearly preferable to use N_o rather than surface values N_s. According to the maps, monthly mean values of N_o may vary from 400 to 290 within ± 25° latitude from the equator, with somewhat less variation elsewhere. In the UK, values of about 320 and 340 are typical of winter and summer, respectively. The least-negative mean value of ΔN, about −30, occurs very locally over land at about ± 25° latitude, and the most negative value, about − 100, occurs only locally in the Persian Gulf in August. Off the coast of West Africa a value of − 90 is reached for much of the year. In many temperate climates ΔN remains fairly constant throughout the year, being about − 40 for the UK. The authors of the charts have emphasised that the variation from year to year of mean values of N_s and ΔN for a particular month may exceed the seasonal variability. They also state that the charts are not suitable for mountainous regions or near coasts, especially where the horizontal variation is large. World maps of the mean refractive-index gradient over the first 100 m not exceeded for 10% and 2% of the time have also been published.[13] Atmospheric ducts may occur within this height interval for small time percentages (see Section 2.5).

2.4 Refraction

A general consequence of the atmospheric refractive index changing with height is that, in terms of geometrical optics, radiowaves do not propagate in straight lines. For a spherically-stratified medium, Snell's law may be expressed in polar co-ordinates as

$$n(h)(h + a) \cos \alpha(h) = k \tag{2.21}$$

where $n(h)$ is the refractive index at height h above the earth's surface (radius a), $\alpha(h)$ is the ray angle with respect to the horizontal and k is a constant along a ray. It may be shown that for a vertical gradient of refractive index of dn/dh, the rays are refracted towards the region of higher refractive index with a radius of curvature r, so that

$$\frac{1}{r} = -\frac{1}{n}\frac{dn}{dh} \cos \alpha \tag{2.22}$$

When the refractive index gradient may be assumed constant over a considerable height interval, it is often convenient to use a geometric transformation to produce models for which either straight rays propagate above a model earth of *effective earth radius* a_e, or, alternatively, rays of effective ray radius r_e propagate above a *flat earth*, rather than considering the two true radii a and r together. The models are shown in Fig. 2.4. To equate them, the relationship between δh and d must remain the same. It may be shown that for the straight-ray model (Fig. 2.4a)

$$\frac{1}{a_e} = \frac{1}{a} - \frac{1}{r} = \frac{1}{a} + \frac{dn}{dh} = \left(157 + \frac{dN}{dh}\right) \times 10^{-6} \qquad (2.23)$$

for near-grazing incidence ($\alpha = 0$) with $n \simeq 1$. For the flat-earth model (Fig. 2.4c)

$$\frac{1}{r_e} = \frac{1}{a} - \frac{1}{r} = \frac{1}{a} + \frac{dn}{dh} = \left(157 + \frac{dN}{dh}\right) \times 10^{-6} \qquad (2.24)$$

In this notation the effective earth radius a_e and effective radius of ray curvature r_e are both positive when the gradient, dN/dh, is less negative than $1/a_e$, as is normally the case. The reciprocal of the radius of curvature is referred to as the ray curvature.

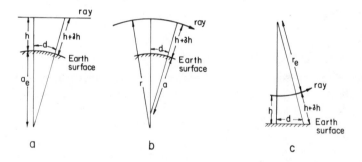

Fig. 2.4. *Comparison of conventional geometric models illustrating ray bending with typical refractive index gradient* ($-157 < dN/dh < 40$ N/km)
  a Straight-ray model
    Effective-earth radius a_e
  b Real earth
    Real earth radius a
    Real ray radius r
  c Flat-earth model
    Effective ray radius r_e

Since atmospheric refractive index normally decreases with height, eqn. 2.24 shows that the effective earth radius is normally greater than the true earth radius. The ratio of the two is referred to as the *effective-earth-radius factor k*, where

$$k = \frac{a_e}{a} = \left(1 + a\frac{dN}{dh} \times 10^{-6}\right)^{-1} \qquad (2.25)$$

(Clearly also $k = r_e/r$). Fig. 2.5 shows the way in which the effective-earth-radius factor k changes with the refractivity gradient dN/dh. A mean refractive index gradient near ground of about -40 N/km, the median value, gives a value of k of $4/3$, i.e. $a_e = ka = 8500$ km. For this reason it is common practice to use a $4/3$ earth radius when drawing to scale the progress of rays from a transmitter. The special case when the ray curvature is the same as the earth curvature occurs when the refractive index gradient is equal to $1/a$, i.e. this gradient is -157 N/km.

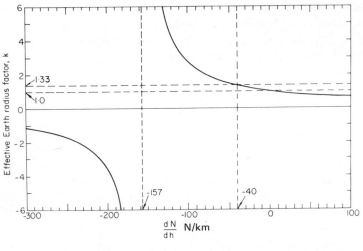

Fig. 2.5. *Effective earth radius factor k as a function of refractive index gradient dN/dh*

Then k, a_e and r_e are all infinite. When the gradient is more negative than -157 N/km, k, a_e and r_e are all negative and δh (shown in Fig. 2.4) decreases with increasing distance (i.e. the ray bending is towards the earth's surface).

By way of common nomenclature, if the ray is bent downward less than the normal, i.e. $dN/dh > -40$ N/km, it is said to be *subrefracted*. If it is bent downward more than the normal, i.e. $dN/dh < -40$ N/km, it is said to be *superrefracted*. If it is bent downwards so as to have a radius of curvature less than that of the earth, the possibility of ducting occurs. Ducting will be considered in the next Section, but it is convenient to introduce here the concept of a modified refractive index M, which is related to N by

$$M = N + \frac{h}{a} \times 10^6 \qquad (2.26)$$

Then

$$\frac{dM}{dh} = \left(\frac{dn}{dh} + \frac{1}{a}\right) \times 10^6 = 10^6/r_e \qquad (2.27)$$

i.e. if dM/dh is constant with height, then the effective ray curvature is constant for a flat earth model. Furthermore

$$\frac{dM}{dh} = \frac{dN}{dh} + 157 \quad \text{N/km} \qquad (2.28)$$

i.e. dM/dh is positive for N gradients less negative than -157 N/km, for which rays are refracted away from the flat earth, and negative for N gradients more negative than -157 N/km, for which refraction is towards the flat earth and which can therefore cause ducting. The gradient of M is consequently a useful indicator as to whether ducting may occur.

Because of refraction, an *elevation angle correction* ϵ is required when directing a ground-based steerable antenna at a target, whether that target is

(*a*) outside the earth's atmosphere, e.g. a satellite
(*b*) within the atmosphere but above the antenna, e.g. an aircraft, or
(*c*) at some distance away horizontally, e.g. a distant transmitter or receiver.

The general case is indicated in Fig. 2.6. Owing to the cosine term in eqn. 2.22 and the fact that less atmosphere is traversed, this correction is small at high elevation angles. However, it may become significant at low elevation, especially if narrow-beam antennae are used, and it is often useful to examine by a *ray-tracing* technique the ray bending due to assumed or measured refractive index gradients. Within the first few kilometres from the ground, where refractive index changes with height are most variable, this is an iterative process of evaluating the radius of curvature r of the rays according to the refractive index gradient dN/dh and elevation angle α in successive height intervals (using eqn. 2.22),

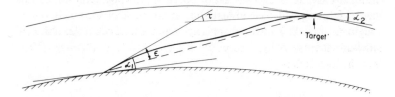

Fig. 2.6.   *Ray bending τ and elevation error ϵ due to refraction*
(True elevation angle of target $= \alpha_1 - \epsilon$)

and so plotting the ray trajectory. The distance travelled in each small height interval δh of a spherically-stratified (or 'horizontally-stratified') medium is given by

$$\delta h = \alpha_1 d + \tfrac{1}{2} d^2 \left(\frac{1}{a} - \frac{1}{r} \right) \tag{2.29}$$

i.e.

$$\delta h = \alpha_1 d + \tfrac{1}{2} d^2 \left(157 + \frac{dN}{dh} \right) \times 10^{-6} \tag{2.30}$$

see Fig. 2.7, and the ray bending τ and elevation angle α_2 at the top of the height interval are given by

$$\tau = \frac{d}{a} + (\alpha_1 - \alpha_2) \tag{2.31}$$

or

$$\tau = \frac{d}{r} = -d \frac{dn}{dh} \tag{2.32}$$

assuming $\cos \alpha \simeq 1$. Although for simple linear or exponential models of the atmosphere there is a clear relationship between ray bending and elevation error;[2] in the general case there is no such relationship.

The viability of the ray-tracing technique is restricted in that

(*a*) the refractive index should not change appreciably in the space of a wavelength and

(*b*) the fractional change in the spacing between neighbouring rays (initially parallel) must be small in a wavelength.

To meet these conditions the refractive index-height profile must not have an abrupt change or too large a gradient. Both conditions are violated when ducting occurs, since this mode of propagation is similar to that of a waveguide, but ray tracing may still give some useful insight. It is also useful in assessing elevation error corrections due to refraction on long radio paths. In more sophisticated ray-tracing programs, the refractive index-height profile varies with distance, although the use of the technique is then usually limited by lack of meteorological data.

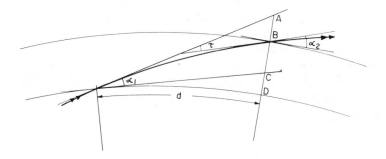

Fig. 2.7. *Ray bending in a small height interval* δh *with uniform refractive index gradient*
Ray radius $= r$, earth radius $= a$
$AB \simeq d^2/2r$, $AC \simeq d\alpha_1$, $BD = \delta h$, $CD \simeq d^2/2a$

Statistical information on the mean ray bending for a radio path through the whole atmosphere, and the day-to-day and short-term variations of this ray bending, are presented in Section 5.2. The ray bending and elevation corrections are then identical. Just as the atmospheric refractive index is substantially independent of the radio frequency, so too is the ray bending.

2.5 Ducting

In meteorological conditions where over a large horizontal area the atmospheric refractive index decreases sharply with height, radiowaves can be trapped and experience low-loss propagation over long distances. This condition is known as 'tropospheric ducting'. Although it is a frequent phenomena in some places and under certain meteorological conditions, it is not a sufficiently reliable mode for communication purposes. However, it can cause strong interference fields well beyond the horizon (see Section 8.3.2), and can also cause severe fading on line-of-sight paths (see Section 4.3).

The first necessary condition for a duct to occur is that the refractive index gradient shall be *equal to or more negative than* − *157N/km.* This causes the rays to remain close to the earth's surface beyond the normal horizon (even if the refractive index gradient is not quite sufficiently negative to produce a duct, some radio energy will pass beyond the normal radio horizon).

The second necessary condition, that this gradient should be maintained over *a height of many wavelengths,* has been examined in terms of a mode theory.[19] Although natural ducts do not have the sharp boundaries of a metallic waveguide, they do have a wavelength cut-off above which waves will not propagate. Since the duct thickness t(m) does not have a sharp limit, so the cut-off wavelength $\lambda\,max$(m) is not sharp, but an indication of the relationship between them is given by

$$\lambda\,max \ = \ 2{\cdot}5 \times 10^{-3} \left(\frac{\delta N}{t} - 0{\cdot}157\right)^{1/2} t^{3/2} \qquad (2.33)$$

where δN is the refractive index change across the duct. By way of example, a typical duct near the ground with a thickness of 25 m and refractive index change of 10 N-units (i.e. a gradient of − 400 N/km), the cut-off wavelength is 0·15 m, whereas a duct of the same gradient would have to extend over a height interval of about 87 m to propagate a wavelength of 1 m. Normal duct thicknesses are such that complete trapping occurs only at SHF, and only in extreme conditions does complete trapping occur at VHF. Although ray theory cannot give as full a description of ducting as can mode theory, it is a useful means of first visualising the nature of ducts.

For reasons to be considered shortly, ducts more frequently occur near to the ground level. Figs. 2.8a to d show rays propagating within ground-based ducts according to ray theory. The duct profiles are drawn in terms of M rather than N because (as was mentioned in Section 2.4) it is useful to employ the sign of the gradient dM/dh as in indicator of whether ducting can occur. For zero M gradient, rays leaving the transmitter at elevation angles close to the horizontal will travel parallel to the earth's surface, while rays at other elevation angles will continuously travel away from the earth or strike the ground. If the M gradient is negative and uniform within the duct, rays starting at elevation angles below a critical value θ_c will strike the ground, and those with initial elevation angles greater than θ_c will leave the duct (see Fig. 2.8a). It may be shown that

$$\theta_c = (2\Delta M \times 10^{-6})^{1/2} \qquad (2.34)$$

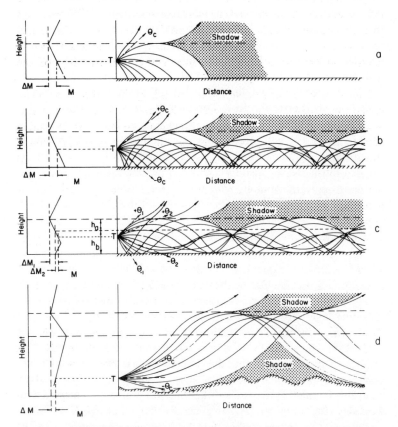

Fig. 2.8. *Rays propagating from a transmitter situated in a ground-based duct*

  a Uniform $dM/dh < 0$ in duct ($dN/dh < -157$ N/km) and $dM/dh > 0$ above duct ($dN/dh > -157$ N/km). Rays trapped for $\theta < \theta_c = \sqrt{(2\Delta M \times 10^{-6})}$

  b As for *a*, but with ground reflection

  c As for *a*, but with $dM/dh > 0$ near ground. If transmitter is in height region h_a, rays remain in duct for θ between $\pm\theta_1 = \pm\sqrt{(2\Delta M_1 \times 10^{-6})}$ as for *a* and *b* above. If transmitter is in height region h_b, rays remain in duct without ground reflection for θ between $\pm\theta_2 = \pm\sqrt{(2\Delta M_2 \times 10^{-6})}$ and within the duct but with ground reflection if θ between $\pm\theta_1 = \pm\sqrt{(2\Delta M_1 \times 10^{-6})}$

  d A special case of *c* above. Rays trapped for θ between $\pm\theta_c = \pm\sqrt{(2\Delta M \times 10^{-6})}$

where ΔM is the increase in the modified refractive index at the transmitter height compared with that at the duct top. In practice the value of θ_c never exceeds about $0 \cdot 5°$ (which corresponds to $\Delta M = 38$, e.g. a decrease of 54 N in 100 m or 195 N in 1 km). In the absence of ground reflection, there will be a shadow region, as shown in the Figure, for which diffraction theory is not applicable. If, however, the surface reflectivity is high, as over sea, continued surface reflections carry rays launched within an angular range $\pm \theta_c$ far beyond the normal horizon, as shown in Fig. 2.8b. Normally the refractive index gradient will not be constant within the duct. If it has the form shown in Fig. 2.8c, energy may be ducted for large distances even if the ground reflection coefficient is poor, but only for a low transmitter. If a transmitter is situated in the height region h_a, then rays within $\pm \theta_1 = \pm (2\Delta M_1 \times 10^{-6})^{1/2}$ will propagate within the duct with ray bounces in the manner considered for Fig. 2.8b. Similarly, for a transmitter situated in the height region h_b, this will be true for rays within $\pm \theta_1$. However, rays transmitted within $\pm \theta_2 = \pm (2\Delta M_2 \times 10^{-6})^{1/2}$, ΔM_2 being the difference in M value between the transmitter height and ground level, will be confined within the height range h_b and will propagate without ground reflection (and the associated losses).

Fig. 2.8d shows a rather special case of ground-based ducting, in which the top of the duct is well above ground level and the downwards refraction all occurs close to the top of the duct. For long wavelengths or a very abrupt refractive index change, this elevated region may act as a reflecting layer (see Section 2.6). In a particular case studied at VHF, radio energy was carried by this means well beyond the normal horizon in a region where terrain roughness would limit low-level ducting.[20] A ray-tracing study showed that the refractive index variation with height was such as to produce a coherent phase front at the receiver, and the very slow fading that occurred at the time was probably associated with focussing effects as the refractive index distribution changed.

Because the energy within the duct spreads with distance in the horizontal, but is constrained in the vertical direction, it is possible in principle that the field strength within a duct may be greater than the free-space field strength for the same distance. With no spread in the vertical but normal spread in the horizontal, the transmission loss might be expected to be proportional to the distance d, instead of following the free-space d^{-2} law (as discussed later in this Section). However, the duct will normally be '*leaky*' i.e. some energy will steadily pass out of the top of the duct, thereby adding to the transmission loss within the duct. A consequence of this leakage is that the field strength just above a duct at a distance well beyond the normal horizon may be higher than were the duct not present, and, by reciprocity, the signal level within the duct would be higher than normal even if the transmitter were just outside the duct. Conversely, within line-of-sight, the shadow effects shown in Fig. 2.8 may lead to much lower signal strength above the duct top if the transmitter is within the duct, or much lower signal strength within the duct if the transmitter is above. Both these factors have practical significance, not least in the operation of radars with beams close to horizontal over sea, where 'ground-based' ducting is most likely to occur. For instance, if a sea-area-surveillance radar is placed on a cliff top to obtain long-distance coverage in normal conditions, a low-level duct may considerably reduce the echoes from relatively close objects due to the shadow effect, but enhance echoes from far beyond the normal horizon by duct leakage. Normally, the refractive index gradient will not be constant with horizontal distance, and so a duct will have horizontal limits beyond which the radio energy reduces rapidly. Also, the transmission loss within a duct will change considerably as a function of time as the duct characteristics change.

Ground-based ducts may be formed by an unusually rapid decrease of water vapour with height or an increase in temperature with height or both effects together. Two causes which are associated with the sea or large areas of water are evaporation and advection. *Evaporation* of water vapour from the surface of the sea may cause a zone of high humidity (i.e. high refractive index) below a region of drier air. Such ducts are particularly likely to occur in the afternoon due to prolonged solar heating. The duct thickness is typically 15 m. Over tropical seas, the high humidity existing near the surface produces almost permanent ducts which may contain a change of some 40 N-units. *Advection*, the movement of one air type over another, may cause hot dry air (from the land) to be blown over cold wet air, producing a region of low refractive index above a region of high refractive index. This is most marked at evening with the onset of a land breeze. The duct thickness is typically 25 m. Such a duct may also form when warm dry air is blown over cold ground. The occurrence of evaporation and advection ducts over sea has been mapped over much of the world.[21]

Radiation cooling may also produce temperature gradients which cause ground-based ducting. This occurs when the ground cools at night due to the absence of cloud cover. Air next to the ground becomes colder than that higher up, and the process continues as shown in parts *a* to *d* of Fig. 2.9 as the ground continues cooling. The duct becomes thicker as the night continues. This phenomenon is fairly commonplace in desert and tropical climates. Forecasting of such duct formation

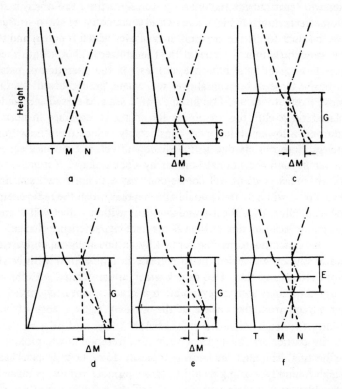

Fig. 2.9. *Profiles of temperature T refractive index N and modified refractive index M as ground-based ducts and elevated ducts form by thermal radiation processes*

Progressive night cooling of ground (*a* to *d*) produces ground-based duct (G). Subsequent solar heating of ground (*e* and *f*) produces elevated duct (E).

———— T
– – – – N
–·–·– M

is difficult because it depends on a number of factors. The process requires a slight breeze to be present to mix the air close to the ground with that slightly higher up. If the air temperature closest to ground level is below its dew point, then the formation of low-level fog or mist may actually reduce the refractive index value since the contribution to the refractive index of the water in droplet form is substantially less than when it is in the form of vapour. The total amount of water in the air remains essentially the same. The rapid increase in refractive index with height close to the ground may cause subrefraction or reflection of radio energy according to the magnitude of the refractive index change, the height over which it occurs and the radio wavelength. This effect was mentioned in relation to part *c* of Fig. 2.8. Fog close to ground level may fill in surface irregularities, thereby improving the surface reflection. It should be mentioned that similar subrefractive or reflective effects may occur if warm moist air is advected over cold ground. If there is no air movement, no duct will form, since the temperature gradient inhibits any convective mixing of air, but subrefraction or reflection are still possible. If the breeze is too strong, the mixing is too thorough for ducting or subrefraction.

Parts *e* and *f* of Fig. 2.9 show what may happen as solar heating warms the air next to the ground in the morning. The remaining region of rapid decrease of refractive index with height now forms an *elevated duct*. However, ducts formed in such a way will usually be short lived owing to convective activity associated with ground heating physically destroying the stable layer. Ducts similar to that shown in Figs. 2.9*e* and *f*, but caused by subsidence, may last for several days. This effect is associated with anticyclonic and generally settled weather conditions. Hot air rises at the centre of the anticyclone and then spreads out horizontally, cools and slowly descends, producing a distinct boundary with the slightly colder air near the earth's surface. The slight rise of temperature with increasing height at the layer boundary is called a subsidence inversion (of the temperature gradient). It may or may not be accompanied by an inversion of the modified refractive index gradient dM/dh, i.e. dN/dh may not be less than -157 N/km. However, as the mass of rising air first cools prior to subsidence, it may lose much of its moisture, and therefore be relatively dry above the colder air once subsidence occurs. The cooling may take place near the adiabatic lapse-rate for saturated air (about $5°$ C/km), and the warming may be near the adiabatic lapse-rate for dry air (about $10°$ C/km). Strictly this should be referred to as a 'pseudo-adiabatic process', since the air acquires the latent heat of condensation as the water drops out. The relative humidity may fall below 10% at about 1 to 2 km height after prolonged subsidence. The temperature and humidity changes may (together or separately) cause an increase in refractive index within a few tens of metres height interval, which extends horizontally over tens or hundreds of kilometres.

Fig. 2.10 shows an elevated duct similar to that of Fig. 2.9*f* but with the addition of rays leaving a transmitter well below the duct. The only way in which energy can get in and out is by the duct acting as a 'leaky' waveguide. Energy leaks in from near the transmitter, and leaks out progressively along the duct with the result that, at a distance, higher signal levels may be received than in the absence of the duct.

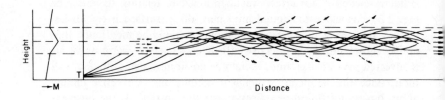

Fig. 2.10.   *Rays propagating from a transmitter situated below an elevated duct*

Similar principles apply for coupling into the duct from above. The calm conditions associated with subsidence-induced elevated ducts may cause ground-based ducts to be present at the same time. Although elevated ducts are less likely to be disturbed by air movements over undulating terrain than are ground-based ducts, the latter undoubtedly play a more major part in producing high signal levels well beyond the horizon, particularly over sea or over large river estuaries or other flat land. Numerous examples have been reported of such ducts causing interference from distant transmissions and producing radar echoes over sea from many hundreds of kilometres.

These illustrations in terms of ray theory give an idea of what may occur for very short wavelengths compared to the duct thickness, but *mode theory* should be used to obtain an understanding of the variation of field strength within and above the duct. In particular, the field in 'shadow' regions will be greater than zero, and the 'leaky duct concept' is somewhat imprecise. Full-wave solutions have been evaluated[22, 23] and these quantify the leakage both above and below the simple uniform duct, and, by reciprocity, the trapping within the duct of radio energy originating from transmitters above or below the duct.

In general terms, the basic transmission loss L_b between two terminals immersed within the duct has been shown to be of the form[24, 319]

$$L_b = 32.4 + 20 \log f + 10 \log d + C_1 d + L_c \qquad \text{dB} \quad (2.35)$$

where C_1 is a constant and L_c is a coupling loss, to be discussed below, and f and d are expressed in megahertz and kilometres, respectively.

Comparing eqns. 1.4 and 2.35, the *path attenuation A* (dB) with respect to free space is

$$A = L_b - L_{bf} = C_1 d - 10 \log d + L_c \qquad (2.36)$$

Whether this attentuation is positive or negative (i.e. whether enhancement with respect to the free-space value is possible) depends largely on the coupling loss term.

When both the transmitter and receiver are within the duct, losses have been determined theoretically to be about $0 \cdot 1$ dB/km for the first four modes of propagation.[25] It has also been shown that the specific attentuation for the first mode is about $0 \cdot 03$ dB/km so long as the wavelength is less than $0 \cdot 0002 t^{1 \cdot 8}$ m, where t(m) is the duct thickness.[23, 25] This gives a lower limit for the term C_1 in eqns. 2.35 and 2.36. For terminals within the duct, the term L_c is determined by the limited angle $\Delta\theta = 2\theta_c$ in Fig. 2.8, being less than the half-power-point antenna beamwidth θ_B. This is analogous to the aperture-to-medium coupling loss (gain degradation) of antennae considered in Section 6.2. From these considerations[24]

$$L_c = 10 \log (\Delta\theta/\theta_B) \quad \text{dB} \qquad (2.37)$$

for $\Delta\theta < \theta_B$, and zero for $\Delta\theta \geqslant \theta_B$.

When one or both terminals are above or below the duct, the coupling loss will be considerably more than that indicated above. Full-wave solutions for elevated, but non-uniform, ducts[26] can produce computed values of coupling loss (about 10 dB) and transmission loss consistent with measured data (in which VHF transmissions have been observed within 21 dB of the free-space level at distances up to 650 km[27])

2.6 Reflection at refractive index boundaries

If the refractive index gradient at a horizontal boundary between two air masses is sufficiently abrupt compared with the radio wavelength, it may cause partial reflection of radio energy. Whether a horizontally-stratified refractive index gradient more negative than -157 N/km will cause reflection or ducting depends largely on the radio wavelength. Such a gradient might have inadequate vertical extent to cause ducting of the longer wavelengths (see eqn. 2.33), but adequate refractive index change to cause reflection. For this reason ducting tends to be of less importance below about 500 MHz, while reflection tends to be less important above about 1 GHz.

Layer boundary gradients may be enhanced if the rate of decrease of temperature with height (lapse rate) below the boundary is greater than the adiabatic temperature lapse rate. Then a parcel of air at ground which is slightly warmer and/or more moist than its surrounding air, will rise from the ground and continue to do so until it meets the temperature inversion layer. When introducing eqn. 2.6, mention was made that moist air has a lower density than dry air at the same temperature. As the parcel of air ascends, it continues to be warmer than its surrounding air, since its rate of cooling due to adiabatic expansion is less than the rate of decrease of temperature with height of the ambient air. By this transport process the temperature inversion may cause a layered humidity change, and the two effects together may produce a very pronounced refractive index change over a large horizontal extent.

A statistical study of the maximum refractive index gradient in any 75 m height interval within the first 1200 m above ground in central England[12] showed no prevalent height for the maximum (apart from the first 75 m). It also showed that, under median conditions, the contribution to this refractive index lapse-rate due to temperature was zero, while that due to humidity was 63%. When the gradient was less than -120 N/km, the contribution due to humidity was 75% of the total. It was more than half the total on 70% of occasions. Seldom was the temperature contribution dominant. Natural processes may be expected to neutralise localised differences of temperature more rapidly than they will neutralise localised changes of humidity. A study in the UK of the heights at which layers were observed with a radar has shown the most common to be at about 1·4 km.[28] However, measurements with microwave refractometers have shown refractive index fluctuations to decrease with height above ground.[29, 30]

For continuous single boundaries between two air masses, the evaluation of the reflected energy at a distance depends on

(a) the reflection coefficient ρ for a smooth plane of infinite extent having the same refractive index change
(b) on any surface roughness of the boundary
(c) on any general curvature of the boundary and
(d) on the horizontal area of the boundary.

The approach is similar to that for calculating contributions from ground reflection as discussed in some detail in Section 4.2.3. The reflection coefficient ρ for a plane surface of infinite horizontal extent depends on

(i) the change of refractive index Δn at the boundary
(ii) the height interval over which it occurs Δh

(iii) the form of the way in which the change occurs and
(iv) the grazing angle α of the incident energy (i.e. the angle of elevation at a horizontal layer).

For an abrupt difference of refractive index between two air masses (i.e. $\Delta h = 0$) there will, in general, be refraction of energy passing the boundary, and partial reflection at the boundary, with the magnitude of the reflection coefficient $|\rho_0|$ given by Fresnel's formula

$$|\rho_0| = \frac{\alpha - (\alpha^2 - 2\Delta n)^{1/2}}{\alpha + (\alpha^2 - 2\Delta n)^{1/2}} \tag{2.38}$$

for small angles α, so that $\alpha \simeq \sin \alpha$. This reduces to

$$|\rho_0| = \frac{\Delta n}{2\alpha^2} \tag{2.39}$$

if $2\Delta n \ll \alpha^2$. The reflection coefficient rises to unity for energy incident at a grazing angle less than a certain value α_c on the high refractive index side of a boundary where there is an abrupt change with height Δn and where $\alpha_c = (2\Delta n)^{1/2}$. This condition is referred to as '*total internal reflection*', and the complement of α_c (i.e. $(\pi/2) - \alpha_c$) is called the 'critical angle'. Because α is small, there is also a phase change of π.

In general, the refractive index change will not be abrupt, and consequently the reflection coefficient of eqn. 2.39 will be reduced due to a '*form function*' F_ρ, which depends on the way in which the refractive index changes, i.e.

$$|\rho| = |\rho_0| F_\rho \tag{2.40}$$

This form function may be calculated for simple models of the boundary form.[31] For instance, for a linear change in gradient $\Delta n/\Delta h$, over a height Δh, as indicated in Fig. 2.11, with smoothing over a height interval δ at top and bottom of the layer, it can be shown that[31, 32]

$$F_\rho = \frac{\sin (4\pi\alpha\Delta h/\lambda)}{4\pi\alpha\Delta h/\lambda} \tag{2.41}$$

so long as $\Delta n \ll \pi\alpha^3\Delta h/\lambda$, and $\delta \ll \lambda/(4\alpha)$. For any of these models, if the boundary is abrupt, i.e. $\Delta h/\lambda$ is small and/or α is small, then the form function becomes unity, and the reflection coefficient is $|\rho_0|$, as given by Fresnel's formula. No model of this type will describe all reflecting layers, but reasonable confirmation of the validity of the models has been obtained from meteorological measurements made at the centre of VHF radio paths.[20] As with other aspects of radio-meteorology, agreement between theory and practice is limited because only incomplete meteorological data can be obtained on a radio path. Furthermore, there will be differences in layer characteristics with distance.

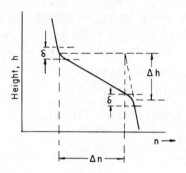

Fig. 2.11. *Linear refractive index-height profile at layered air mass boundary, with transitional height zones* δ

Because the magnitude of F_ρ is inversely proportional to $\Delta h/\lambda$, layer reflection is not of much consequence above UHF. However, at VHF it may cause considerable interference problems on transhorizon paths and multiple-path interference on line-of-sight paths (as discussed in Sections 8.3.2 and 4.4., respectively). Such a boundary may extend for many tens of kilometres in horizontal extent at a height of one or two kilometres, with a 20 to 30 N-unit refractive index change in a vertical depth of only 30 to 50 m. When reflection is from a surface whose dimensions are small compared with the first Fresnel zone (see Section 4.2.3); i.e. for dimensions $l_1 < \sqrt{(d\lambda)}$ across the path and $l_2 < \sqrt{(d\lambda)}/\alpha$ along the path of length d, the reflection coefficient is reduced by a factor $4l_1 l_2 \alpha/\lambda d$.[32] For the most common case, $\sqrt{(d\lambda)} < l_2 < \sqrt{(d\lambda)}/\alpha$, the reduction factor is $2l_2\alpha/\sqrt{(d\lambda)}$. For a 'troposcatter' path extending well beyond the radio horizon, there may be a large number N of these reflecting surfaces in the volume V of atmosphere common to the transmitting and receiving beams, in which case the equivalent reflection coefficient will be

$$\left[\frac{2N}{d\lambda} \int |\rho_0|^2 F_\rho^2 l_2^2 \alpha^2 \, dV \right]^{1/2} \tag{2.42}$$

Although the reflection coefficient at a refractive index discontinuity layer boundary is usually much less than unity, a study of VHF data obtained on two paths of about 300 km length has established that *double-hop reflection* from low-level layers may produce a comparable transmission loss with that of single-hop reflection from similar layers.[33] This happens because the reflection coefficient is an inverse function of a power of the grazing angle, and this angle for a single reflection α_1 is twice that for double-hop reflection α_2 in the idealised case shown in Fig. 2.12. However, the numerical effect of reducing the grazing angle is not only

(*a*) to increase the reflection coefficient for the equivalent plane surface, but also

(*b*) to reduce the loss due to any variations in the height of the reflecting layer (i.e. the surface roughness factor of eqn. 4.8) and

(*c*) to increase the focussing effect of the concave surface (i.e. the 'divergence factor' *D*, of eqn. 4.14 is in this case greater than unity and can produce a 3 dB enhancement).

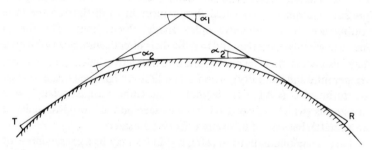

Fig. 2.12. *Path geometry for double-hop and single-hop reflection*

Double-hop reflection is not likely to occur on paths of less than about 200 km length over irregular terrain, since the necessary layer height would then be too close to the hill tops.

During the last decade there has been some study of the '*whispering-gallery*' *mode* of propagation.[34] This phenomenon is similar to the propagation of acoustic waves along a concave wall, since low-attenuation propagation may occur in the concave boundary between horizontally-stratified air masses. As with the case of elevated ducts, such modes are not effectively excited unless the source is situated in the neighbourhood of (or above) the concave boundary, but recent work suggests that energy enters and leaves these 'whispering-gallery' modes at lateral discontinuities in the path of the wave.[35] Here conversion takes place between the modes that extend up to the layer boundary (but are highly attenuated with distance) and the modes within the boundary (that have low attenuation).

2.7 Scatter from refractive index fluctuations

So far this Chapter has been concerned with large-scale height gradients of refractive index. The present Section considers small-scale variations of refractive index due to local pockets of high or low temperature and/or humidity. These have scale sizes too small for their effects to be treated adequately on a simple refraction basis. If the refractive index irregularities are of sufficient intensity, they will scatter radio energy, and the amount of scattering in a given direction will depend on the spatial distribution of the irregularities and the radio wavelength. The irregularities will also cause scintillation of a radiowave passing through (see Sections 4.5 and 5.3.2). Pressure fluctuations are rarely important for the size and time scales of interest here since localised changes in pressure are quickly dissipated. It is important to distinguish the term 'turbulence' from 'refractive index spatial flucutations'. The former may accompany the latter, and may be their direct cause, but turbulence may also exist in a region of uniform refractive index (i.e. uniform temperature and humidity). A considerable study has been made of clear air turbulence (CAT), as distinct from turbulence associated with rainstorms (which, unlike CAT, may be seen on the low-power radar of an aircraft), because of its severe effect on air safety.

Large-scale fluctuations of refractive index may be a consequence of

(*a*) the horizontal movement of one air stream over another
(*b*) vertical movement of moist and/or warm air parcels (see Section 2.6) or
(*c*) any other process of entrainment whereby parcels of one air mass are injected bodily into a neighbouring region of different temperature or humidity.

As a result of relative movement of two air masses, some of each air type (wet, dry, cold or warm) is dragged in towards the other. The parcels of air of differing refractive index are then further broken down by the straining of large-scale inhomogenities by wind shear, and the mixing of an otherwise stable inhomogeneous distribution by turbulence. As the process proceeds and the mixing becomes more complete, smaller and smaller scale 'blobs' of dissimilar air will be generated, the smallest blobs being dispersed by eddy diffusion. So long as the two main air masses continue their relative movement, the creation of large blobs and their subsequent breakdown continues, and a spectrum of blob sizes exists at a given time.

A *description of the spatial variations* of refractive index fluctuations, as would be measured along a line in space, can be given

(*a*) in terms a structure function $D_n(r)$, where r is the point separation, or

(*b*) in terms of a one-dimensional space spectral function $F_n(k)$, where k is the wave number corresponding to the spatial scale $l = 2\pi/k$.

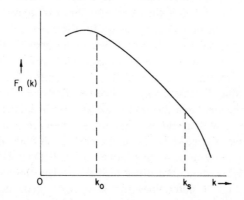

Fig. 2.13.   *Spectrum of refractive index irregularities plotted versus wave-number showing various turbulence ranges*
  $0 < k < k_0$:   Input range
  $k_0 < k < k_s$:   Inertial subrange or transformation range
  $k_s < k < \infty$:   Dissipation range

For a fully-developed turbulent process, the spectral function has the form of Fig. 2.13. The 'input' or 'blob-creating' range $0 < k < k_0$ corresponds to very large blobs which are normally anisotropic, and the scale length $l_0 = 2\pi/k_0$ (i.e. the average size of the blobs in the hierarchy of decaying eddies) is of the order of 100 m. For very small blobs, the 'dissipation range' $k_s < k < \infty$ is characterised by a sharp drop in turbulent activity and in the spectral intensity owing to the destructive action of viscosity and diffusion. The scale length $l_s = 2\pi/k_s$ is of the order of 1 mm. The intermediate 'inertial subrange' $k_0 < k < k_s$ is characterised by the process of redistributing the turbulent energy towards higher wave numbers and represents the progressive subdivision of eddies in the turbulent structure. Tropospheric scatter propagation depends principally on this inertial subrange, since it is within this range that the blob sizes are comparable with the wavelength.

If within the inertial subrange the refractive index fluctuations are isotropic, then within a homogeneous region[36]

$$D_n(r) = C_n^2 r^{2/3} \tag{2.43}$$

and

$$F_n(k) \simeq \tfrac{1}{4} C_n^2 k^{-5/3} \tag{2.44}$$

where C_n^2 is the structure constant.[37] C_n^2 is a measure of the variability of the refractive index fluctuations. Values near the surface are typically 10^{-12} to 10^{-14} m$^{-2/3}$, but they decrease abruptly above about 2 km height.[38] Attempts to measure the spectral form of refractive index fluctuations rely on Taylor's hypothesis of a 'frozen' structure. By this means the measured time series may be converted to a spatial representation by assuming a velocity of movement but no internal change with time. It is necessary to ascertain that the fluctuations are isotropic for the measured spectra to have any useful application.

Measurements with a balloon-borne refractometer in or near elevated inversion layers of large refractive index variance, correlated with radar echoes, have given values of the exponent m in $F_n(k) \propto k^{-m}$ generally greater than 5/3 and occasionally as high as 3.[39] The importance of this exponent is that it determines the wavelength dependence of the *scattering cross-section per unit volume* η, i.e. the total scattered power per unit volume per unit incident power density. Theory for isotropic turbulence in an inertial subrange predicts that $\eta \propto \lambda^{-1/3}$, but experiments show that the mean value of the wavelength dependence is given by $\eta \propto \lambda$. All theories based on atmospheric turbulence result in an expression of the form

$$\eta = \frac{\pi(\Delta n)^2 \sin^2(\beta) K^2 F_n(K)}{8 \sin^4(\theta/2)} \tag{2.45}$$

where $F_n(K)$ is evaluated at $K = (4\pi/\lambda) \sin \theta/2$, θ is the scattering angle, β the angle between the direction of scattering and the electric field in the incident wave and $\overline{(\Delta n)^2}$ is the total variance of the refractive index fluctuations. This may be used in the general scatter equation:

$$\frac{P_r}{P_t} = \frac{G_t}{4\pi d_t^2} \frac{\sigma}{4\pi d_r^2} \frac{G_r \lambda^2}{4\pi} \tag{2.46}$$

where

$$\sigma = \eta V \tag{2.47}$$

and where P_r and P_t are the received and transmitted powers, G_r and G_t are the gains of the antennae which are at distances d_r and d_t from the scattering volume V, and λ is the radio wavelength. For the commonly-used frequency range of 300 to 3000 MHz, and scattering angles of 0.7 to $3°$ (i.e. ranges of 200 to 900 km, assuming horizontal beams at the path terminals), the scale length $l = 2\pi/k = \lambda/(2 \sin \theta/2) \simeq \lambda/\theta$ lies between 2 and 80 m, which is the inertial subrange. For studies by radar, the relevant scale length is $l = \lambda/2$, so that centimetric-wavelength radars may be used to examine the inertial subrange.

However, an important fact that has been established by carrying a refractometer on a captive balloon is that the regions of large refractive index variance are generally localised to horizontally-stratified height intervals several tens of metres thick close to intervals having temperature increase with height or without temperature change.[28, 39] These measurements, together with earlier ones which included the use of a vertically-pointing radar,[40] established that the radar echoes and regions of high refractive index variance were often associated with marked wind shear, but this was not always the case. As well as being localised in a restricted height interval, the refractive index fluctuations were localised horizontally. This intermittency or 'patchiness', of the scattering elements is of considerable significance, since it means that they will fill only a part of the atmospheric volume common to the transmitting and receiving antennae of a scatter link. Even for antenna beams no more than $1°$ wide, the common volume of a typical forward-scatter link extends over a vertical distance of $1-2$ km and a horizontal distance of $20-30$ km along the path. Consequently, it must be assumed that the propagation mechanism on transhorizon links is due to scattering from localised patches or layers of large scattering cross-section, rather than bulk scattering from a homogeneously-filled common volume. This conclusion is supported by earlier results of beam-swinging and spaced-antenna experiments carried out in the USA[41] and UK[42], and it may be the prime reason why it is difficult to reconcile the observed results of transhorizon radio transmission with theory.

As well as the non-uniformity of the refractive index fluctuations, there is some question as to their *isotropy*. Wherever a parcel of air is subjected to wind shear without major turbulence, it is to be expected that the orientation of the parcel will be altered and its shape distorted in such a way as to intensify local gradients. By this means, vertical shear of a horizontal wind may cause larger local gradients of refractive index in the vertical than in the horizontal, an effect which should enhance forward scatter. However, radar and refractometer studies have not produced convincing evidence of marked nonisotropy as a normal feature.

Because the theory does not give a good explanation of frequency dependence, and because of the lack of information as to how refractive index fluctuations decrease with height (and their great variability with location and time), the theory has very limited applicability so far as prediction techniques are concerned. For this reason, a semi-empirical approach will be presented in Chapter 6 when discussing the reliability of transhorizon scatter links.

Chapter 3

Precipitation, cloud and atmospheric gases

3.1 Introduction

With the increasing need to use the SHF and EHF bands of the radio spectrum, there has been very considerable study in recent years of the effects on radiowaves caused by rain, ice particles, hail, snow, cloud, fog and even the atmospheric gases. Small displacement currents set up in hydrometeors intercepting a radiowave give rise to re-radiation (scattering) and absorption (due to heating) at these wavelengths. Both may contribute to the attentuation on a radio path, while scatter may also cause interference between radio paths. The relative contribution of the two to attenuation by rain depends on the drop sizes compared with wavelength.[43] For wavelengths which are long compared with the drop size, i.e. in the SHF band, attenuation due to absorption will be greater than that due to scatter. Conversely, for wavelengths which are short in relation to drop size, i.e. in and above the EHF band, scatter will predominate.

Fig. 3.1 shows the specific attentuation (in decibels per kilometre) at ground level due to oxygen and to a typical water vapour concentration. It also shows the specific attenuation due to rain of various rainfall rates and fog of certain densities. These are all additive. The curve for oxygen and water vapour shows a series of molecular resonance absorption peaks. For the rainfall curves, a rate of 150 mm/h occurs for only 0·001% in a tropical climate, whereas a rainfall rate of 25 mm/h may occur for 0·02% of time in a temperate climate (see Fig. 3.7). Note that these rainfall rates are the short-term rates measured over a period of seconds or minutes, and not the integrated rainfall over an hour. The effects of fog or cloud are less severe, although for a dense fog (e.g. 1 g/m³ giving a visibility of 50 m) the specific attenuation will generally be greater than that of the atmospheric gases. Compared with this, drizzle of 0·25 mm/h causes less attenuation than dense fog, because the mass of water per unit volume is so much smaller.

Snow and hail have a relatively small effect, because the complex permittivity of ice is so much less than that of water. However, when snow begins to melt as it falls through warmer air, the melting snowflakes may hold water in very large drops compared with normal rain, and so cause very much more scatter or attentuation. At the other extreme, the small drops that occur in clouds or drizzle have little effect on radiowaves of frequencies less than about 20 GHz, but, at higher frequencies, they may cause significant attenuation.

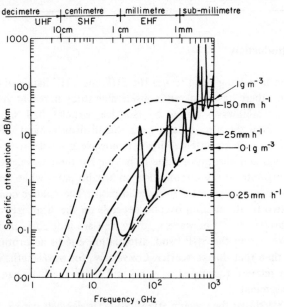

Fig. 3.1.   *Comparison of specific attenuation due to gaseous constituents, fog and precipitation near the surface*
———— Gaseous attentuation calculated assuming 1013 mb, 20° C, 7·5 g/m³ (see Fig. 3.14)
—·—·— Rain attenuation for rates shown (see Fig. 3.12)
———— Fog attenuation for water content shown, at 20° C (see Fig. 3.13)

[Based on CCIR[322]]

3.2   Characteristics of precipitation and cloud

Before examining the effects of rain on radiowaves in detail, it is useful to consider briefly some situations in which rain occurs and the form it takes. Three general classifications of rain that are normally used are 'orographic', 'convectional' and 'cyclonic', but these categories are by no means mutually exclusive.

Orographic rain is caused, entirely or in major part, by the forced uplift of moist air over high ground. The consequent adiabatic cooling of the rising air causes orographic clouds to form, followed, in the event of continued uplift of the air, by precipitation (usually rain) as the air temperature falls below the dew point. The warm sector of a vigorous depression is the synoptic situation in which the orographic influence on rainfall is generally most evident. Even where rainfall is predominantly cyclonic or convectional in nature, the orographic influence is always present to some extent, i.e. greater rainfall rates are recorded on the side of high ground facing the prevailing winds.

Convectional rainfall is caused by the vertical motion of an ascending mass of air which is warmer than its environment. The horizontal dimension of such an air mass is generally of the order of 15 km or less. Convectional rain is normally of greater intensity than orographic rain and is sometimes accompanied by thunder. Because this convective activity is caused by the ground being strongly heated by the sun, it is found in the UK to be most common in mid-afternoon in the months of July and August, although it may also continue into the night and occur at other times of the year. The rain is very intense since it is formed quickly as the convected air rises and cools by adiabatic expansion to a temperature below the dew point. The cells of convectional rain usually form in groups within mesoscale areas of about 1000 km^2. The form of the mesoscale pattern can be greatly influenced by local topographical effects.[44] The lifetime of showers and thunderstorms is longer than that of an individual cell, but the lifetime of a cell can vary from a few minutes up to several hours.[45] Their form may remain constant as they move across terrain, or there may be large fluctuations in rainfall rate while the cells are essentially stationary. Convectional rain usually occurs when wind speeds are low.

Cyclonic rain is caused by the large-scale vertical motion associated with synoptic features such as *depressions and fronts*. Fig. 3.2 shows a simplified section through a warm front followed by a cold front. At a warm front the warm air rises up slowly over the cold air as the two advance across country. The conditions are essentially stable, and, as the warm air rises slowly, horizontally-layered 'stratiform' clouds develop followed by rain. Light rain usually intensifies gradually to become of medium intensity until the front has passed. Conversely, at a cold front the cold air forces in below the warm air, the warm air rises quickly and conditions are unstable. Vertically-shaped pillar clouds develop, and, since the rising air cools quickly, there is a rapid rate of loss of water from the warm air by way of rain.

However, in many instances the horizontal and vertical distribution of frontal rain is much more complicated than Fig. 3.2 suggests.[45-50]

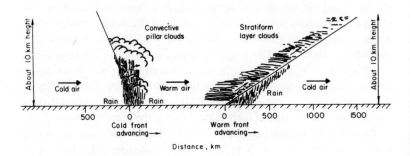

Fig. 3.2.   *Simplified vertical section through warm front and cold front with cloud and rain shown*

The uniform rain mentioned above may extend over a large area (e.g. 200 km × 750 km) parallel to and ahead of the surface position of the warm front (typically 100 km ahead), but in addition there are often bands of heavy rain parallel to and ahead of the surface warm front and also within the warm sector. The latter bands may be oriented at a large angle to the front and extend ahead of the surface position of the front. They may be associated with the local topography. Similar bands may extend for 75 to 150 km ahead of or behind cold fronts. All these bands may be several tens of kilometres wide, with a similar separation, and several hundred kilometres long. They may contain clusters of small-scale convective cells associated with (or triggered by) the frontal ascent. The rain in these convective cells is intense and may be accompanied by thunderstorms. These pockets may have a significant effect on the statistics of attenuation on radio paths. Furthermore, the frontal rainbands sometimes move along their own axes, producing swathes of persistent heavy rain as a frontal system moves through. A radio link oriented along such a swathe may have an interrupted service for a long period of time, while a neighbouring link parallel to the first may be relatively unaffected. This makes for considerable difficulty in predicting the attenuation due to raincells. Generally, the wind speeds, and consequential rates of horizontal movement of rain, are higher for frontal rain than for the convectional rain considered earlier.

Two additional cases of extreme rain conditions are *tropical cyclone storms* (i.e. hurricanes or typhoons) and *monsoon rain*. The former are moving circular storms with intense convective rain of 50–200 km in diameter and extending to 8 km or so vertically. The latter is a very intense form of stratiform rainfall. This type of rainfall, which occurs in the tropics, may last for several hours each day and extend over several hundred kilometres.

Some knowledge of the general characteristics of rainfall to be expected in various locations is necessary if data from one location is to be applied to another. However, when investigating the effects of rainfall on radiowave propagation, the radiometeorologist often makes a simpler breakdown of rain types, distinguishing only between (a) *cells of intense convectional shower-type rain* and (b) *widespread rain from stratiform cloud.* The former may be in cells typically a few kilometres across, and may extend to 10 km in height with rainfall rates of about 85 mm/h, typically exceeded for a total of 0·001% (i.e. 5 min) of an average year at a point on the ground. On the other hand, rain from stratiform (layered) clouds may, in temperate climates, cover regions of some hundreds of kilometres in diameter and extend up to the 0°C isotherm with rainfall rates of 25 mm/h, which is typically exceeded for a total of 0·02% (i.e. 100 min) of an average year. In temperate regions the two rain types may be found in close proximity.

Fig. 3.3 shows two examples where vertical scans of a radar have picked out stratiform rain beneath a highly reflective 'bright band' (the melting layer) at 2 km height. The two vertical sections were obtained in different directions, but each shows a convectional cell apparently standing up through the bright band to a height of 4 km. The bright band is less prominent near the convective cell. The example is taken from a study using a 10 cm radar mounted on a 25 m diameter steerable antenna in Southern England.[51] The good detail of Fig. 3.3 is a consequence of the beamwidth being only 15 arcmins. Notable points are the very rapid changes of rainfall rate with height and distance, and the way that the rain cells in the stratiform area seem to be inclined from the vertical. This inclination may be explained by shear of horizontal wind with height (in both speed and bearing).[51] Since inclination angles of 30° are not uncommon, it is likely that the direction of inclination influences attenuation on ground-to-satellite paths. Subsequent measurements (using a more sophisticated radar system) have shown that cells of high radar reflectivity of the type shown in Fig. 3.3 sometimes contain water drops throughout their height, but sometimes have water drops only below the bright band and ice crystals above.[52]

In some climates, notably Europe and North America, progress has been made in producing *models of the spatial distribution* of the rainfall rate in rain cells using data collected as they passed over a point on the ground. In some of these methods the rainfall rates in the cell are approximated by a log-normal distribution and characterised by the mean and standard deviation,[53-55] while in others the approximation used is that of a gamma distribution.[56] For the conversion of data collected at a point from rain-rate versus time to models of rain-rate versus distance, a good estimate of the velocity of intense rain cells seems to be the wind speed measured at the 700 mb level.[57, 58]

Measurements made with rapid-response rain gauges in the UK have shown that *intensity versus time profiles* of intense convective rain from showers, thunderstorms and the passage of cold fronts may in this location be represented in a statistical form by normalised curves that are independent of the peak rainfall rate (for rates above 20 mm/h).[59]

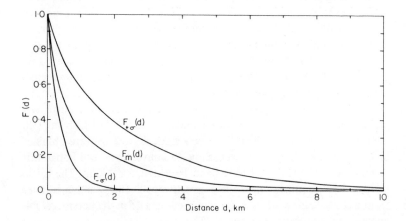

Fig. 3.4.   *Rainfall rates relative to peak value as function of distance from peak for mean and one standard deviation above and below mean, based on UK data*
At distance d approximately 85% of raincells have a rainfall rate less than $F_{+\sigma}(d)$, relative to that at $d = 0$
[Based on data from Harden, Norbury and White[59]]

The intensity versus time profiles were converted to *intensity versus distance profiles* using the wind speed at the 700 mb level. For practical purposes, the experimental intensity versus distance profiles were well fitted by the exponential functions plotted in Fig. 3.4. $F_m(d)$, $F_{+\sigma}(d)$ and $F_{-\sigma}(d)$ are, respectively, the mean intensity and one standard deviation above and below this mean, expressed as fractional values of the peak rainfall rate (i.e. where $d = 0$). In planning studies it is usual to use least-favourable results, and for this purpose it may be noted that, for a given distance from a cell centre, approximately 85% of storms would have rainfall rates less than the top curve in Fig. 3.4 indicates. In particular, the model shows that the rainfall rate of intense rain cells generally falls to about half the peak rate within 1 km of the cell centre. Analysis of the size of intense rain cells in North America tends to support these findings.[58, 60, 61] Although individual storms do not have such an idealised model shape, and are not precisely circularly symmetrical, this model has proved useful for computing attenuation from point rainfall data. It is significant that, although the probability of occurrence of a rain event of a given intensity varies from place to place, the rain cell dimensions seem to be fairly constant in most mid-latitude regions.[44]

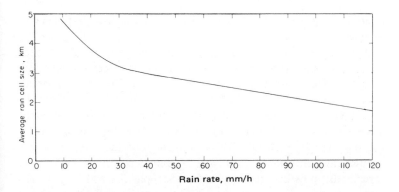

Fig. 3.5.   *Average rain cell size as a function of rain rate*

[Based on CCIR[316]]

Radar studies carried out in Switzerland, Malaysia, Japan and France, have shown the *average horizontal size of rain cells* to be related to rainfall rate (again for rainfall rates above a certain threshold level)[62, 63, 316] in the manner shown in Fig. 3.5, but the form of this curve for a specific locality will depend on the rainfall rate statistics. By contrast, the curves of Fig. 3.4 are less limited by such local variations. Measurements made in Colorado, USA, show that convective showers (which produce heavy rainfall in temperate climates) vary in size during their lifetime and may reach diameters of 15–20 km, although an average size has been estimated as 8 km,[64] which is larger than the cell sizes of Fig. 3.5. Estimation of the *separation between rain cells* is more difficult, but radar measurements made in Canada have indicated an average distance between intense rain cells of approximately 30 km, and the separation distance appears to be governed by a Rayleigh distribution law.[65]

Rainfall rate depends on the *number of drops per unit volume* of air and the statistical *distribution of the drop sizes*. These parameters depend on the rainfall type. A given rainfall rate may result from a lot of small drops or relatively few large drops. Generally, thunderstorms or showers have large drops but fewer per unit volume, whereas in widespread rain or drizzle there will be many more smaller drops. For all but the smallest drops it is found experimentally for various types of rain that the number of drops $N(D)$ of diameter D(mm) per unit volume (m^3) and per unit increment (mm) of drop diameter may generally be approximated by an expression of the form[66, 67]

$$N(D) = N_0 \exp(-D/\bar{D}) \qquad mm^{-1} m^{-3} \qquad (3.1)$$

where $\bar{D}$ is the mean drop diameter. The total number of drops per unit volume is $N_T = N_0 \bar{D}$.

A significant feature of rain drops is that, although small drops fall as spheres, larger drops falling at their terminal velocities will suffer *distortion* due to air drag and may be assumed to be oblate spheroids with a near-vertical minor axis.[68] The ratio of this minor axis a to the major axis b is linearly related to the diameter D_e of the equivolume sphere for $1 < D_e < 10$ mm. To be specific, a/b is 0·97 for $D_e = 1$ mm and 0·41 for $D_e = 10$ mm.[69] This distortion of the drops produces significantly different scatter and attenuation for vertically and horizontally polarised radiowaves (see Sections 3.4 and 3.5). It also contributes to a reduction in cross-polar isolation (see Sections 4.7 and 5.5). These effects of drop distortion may be modified somewhat if the drops have a significant mean canting angle from the vertical or if there is a large spread of canting angles.[70] The former may occur if there is considerable wind shear with height,[71] and the latter may be due to air turbulence associated with wind shear or convection. In practice there is an upper limit to the drop size diameter of about 9 mm, because drops formed with diameters greater than this are so distorted as to be hydronamically unstable, and break-up occurs even if they fall in completely calm air.[68, 69] Radar measurements have confirmed that drops up to this size exist in natural rainfall.[52] This gives a finite upper limit D_{max} for which eqn. 3.1 applies.

Because the drops are generally non-spherical, $\bar{D}$ is not well defined and it is preferable to express eqn. 3.1 in terms of the diameter D_0 of the median equivolume spherical drop, so that $D_0 = 3·67 D_e$ provided that D_{max} is much greater than D_0. This median volume diameter is defined such that half the total water content per unit volume is contained in drops of greater diameter. It has been widely used in preference to D_e since it is more readily measured. The most often quoted drop size distribution model is that of Marshall and Palmer[66] for which $N_0 = 8000 \, \text{mm}^{-1} \, \text{m}^{-3}$ and

$$D_0 = 0.89 R^{0.21} \qquad \text{mm} \qquad (3.2)$$

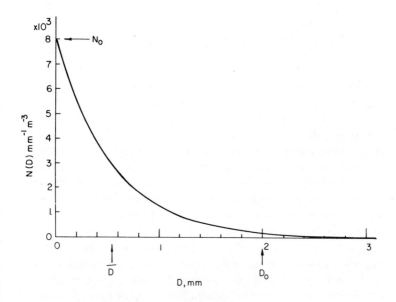

Fig. 3.6. *Example of rain drop size distribution (see eqn. 3.1)* with $\bar{D} = 5.5 \, \text{cm}$, $D_0 = 3.67$, $\bar{D} = 2 \, \text{mm}$, $N_0 = 8 \times 10^3 \, \text{mm}^{-1} \, \text{m}^{-3}$, $N_T = 4 \times 10^4 \, \text{m}^{-3}$ ($R = 47 \, \text{mm/h}$, see eqn. 3.2)

i.e. the diameter of the median volume drop depends on the rainfall rate $R \, (\text{mm/h})$. An example of this drop size distribution is shown in Fig. 3.6, with $D_0 = 2 \, \text{mm}$. Several other exponential drop size distributions have been put forward, and in some of those N_0 is also a function of the rainfall rate.[72]

For a given drop size distribution, the rainfall rate may be determined from the relationship[73]

$$R = 0.6 \times 10^{-3} \pi \int D^3 V(D) N(D) dD \qquad \text{mm/h} \qquad (3.3)$$

where $V(D) \, (\text{m/s})$ is the fall speed of drops of diameter D, and integration is performed over all drop sizes. Alternatively, use may be made of empirical fall speeds,[74-76] and the integral in eqn. 3.3 may then be performed numerically. In this approach it is assumed that the rain drops are falling in still air, whereas, in fact, the rain falling through unit horizontal area in unit time is dependent on the vertical air currents, and these may be very considerable in an intense rainstorm. For this reason, although rainfall rate is a familiar concept, it is sometimes considered preferable to work in terms of the liquid water content (mass per unit volume), also known as water concentration M, rather than the rainfall rate, where

$$M = \frac{\pi}{6000} \int D^3 N(D) dD \qquad \text{g/m}^3 \qquad (3.4)$$

where the density of liquid water may be assumed to be 0.001 g/mm^3. Again, integration is over all drop sizes. Using eqns. 3.1, 3.3 and 3.4, R and M can be expressed in terms of N_0 and D_0. Furthermore, it has been shown that the reflectivity factor Z mm^6 m^{-3} (see eqn. 3.12), the specific attenuation γ (dB/km) at a particular frequency and the optical extinction factor Σ_0 km^{-1} can also be expressed in terms of N_0 and D_0.[73] By establishing any two of these seven parameters, the other five may be determined uniquely. However, as mentioned earlier, Z and γ will depend on the polarisation used, each being larger for horizontal polarisation than for vertical polarisation.

Eqns. 3.1 and 3.2 imply that the probability of occurrence of certain drop sizes depends on the rainfall rate. For very low rainfall rate, i.e. drizzle, most of the drops are between 0.2 and 0.5 mm in diameter, although the limits are not precise. Drizzle forms by coalescence of drops of stratus cloud and falls from the base of this cloud if upcurrents associated with the cloud formation are not too strong. Very small drops are of importance in their attenuation and scattering of sub-millimetre waves. On theoretical grounds, by direct measurement of drop sizes and by the attenuation they produce on submillimetre waves, it has been found that a log-normal drop size distribution gives a better description of these very small drops than does an exponential distribution,[77] but the form of this log-normal distribution is not yet well established.

Clouds consist of a suspension of very small water droplets, ice particles or a mixture of both. At temperatures below $0°$C, cloud particles frequently consist entirely of supercooled water droplets, down to about $-10°$C in the case of layer clouds and to about $-25°$C for convective clouds. At temperatures below these approximate limits, but above about $-40°$C, many clouds are a mixture of ice and water, but with ice crystals predominating at the lower temperatures. Water droplets in cloud generally have a median volume diameter D_0 of about

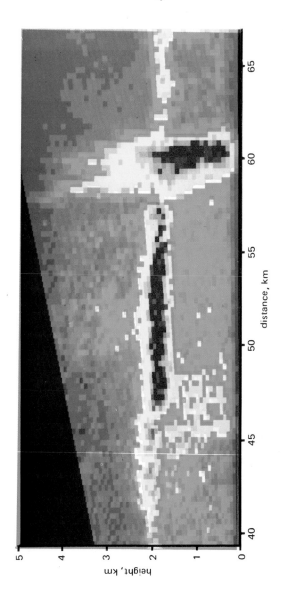

a

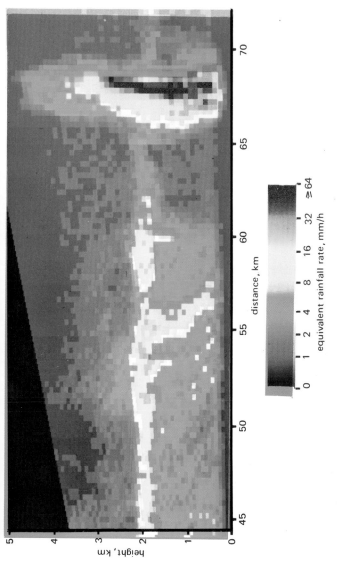

Fig. 3.3. Vertical section through rain cells observed with a 10cm radar
a at 1144 UT   22nd September, 1976
b at 1202 UT

0·015 mm, with a range from about 0·001 to 0·1 mm. Drops larger than about 0·2 mm diameter tend to fall out as drizzle, as mentioned above. Rather larger median values are found in convective clouds (0·015 to 0·02 mm) than in layer clouds (0·01 to 0·015 mm). Measured water-droplet concentrations are generally in the range 10^3 to 4×10^3 m^{-3}, but with smaller values in altocumulus clouds. Ice particles occur in clouds in various forms, determined by such conditions as temperature and degree of supersaturation with respect to ice. Clouds composed of ice particles are very tenuous compared with water clouds. A typical range for the concentration of individual particles is 10^5 to 5×10^5 m^{-3}.[78]

The *liquid water content* of convective cloud is normally computed on the assumption that the water which is condensed on adiabatic expansion is contained within the rising air. The computed maximum values occur towards the top of the cloud and may typically be 5 g/m^3. Because of the great dependence of water content on cloud base temperature and degree of vertical development of cloud, average water-content values of convective cloud are of little significance. Median values of water content of low-level layer clouds (0 to 2 km) in middle latitudes are of the order of 0·2 g/m^3 and about 0·1 g/m^3 for medium level clouds (2 to 7 km). However, values up to about five times the median values have been measured. A convenient assessment of the mean liquid water content of cloud or fog density is in terms of the optical visibility. This has been plotted in Fig. 3.7, based on data put forward by Ryde.[79]

3.3 Measurement and statistics of rainfall rate

It is particularly important in planning telecommunication systems and predicting their performance to have reliable probability-distribution statistics of rainfall rates. Unfortunately, the telecommunications engineer needs them where it is most difficult to be precise, namely, for such small percentages as 0·001% of time (which is 5 min per year). These are difficult to determine for a number of reasons. First, the rainfall rates which occur for very small time percentages may be expected to change very considerably from one year to another, and a very long period (certainly over 10 or 20 years) would be required to define the rainfall characteristics of a particular region. Also there is considerable local variation of the amount of rain that occurs for very small time percentages. The geographical distribution of extreme annual values is greatly influenced by topography, the largest values occurring in hilly regions and the smallest in flat areas. Local topographic features can cause variations in rainfall rate more than an order of magnitude greater than the variations which occur in the absence of any such features.[45] For link planning purposes, it may be necessary to take account of such local variations.

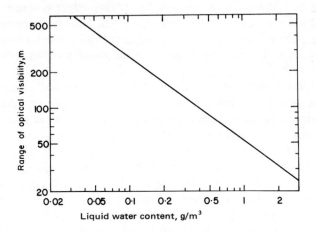

Fig. 3.7.   *Range of optical visibility through fog as a function of liquid water content*

[Based on data from Ryde[79]]

Usually the most intense rain occurs in quite small cells, which pass a point on the ground quickly. Ideally, statistics of the rainfall rate integrated over a minute are required, but the most readily available data are in terms of rainfall integrated over an hour. For some purposes, useful information on the rate of rainfall may be obtained using conventional gauges of relatively slow response, such as natural-syphon recording rain gauges or tipping-bucket rain gauges.[80] However, the most intense rainfall rates, which are particularly important for radiometeorological purposes, occur for such short time intervals that they will not be measured by relatively slow-response instruments. Consequently, several *rapid response gauges* have been specially developed. In one of these instruments the rain is formed into a series of uniform drops, which are then counted to give the flow rate integrated over a few seconds.[81] In another, the rain runs in a narrow channel, the depth of water increases with the flow rate and the capacitance is measured between plates each side of the channel.[82] The prime requirement from studies made with these rapid-response rain gauges is to obtain statistical comparisons with records that have already been obtained using slower-response gauges. The previously recorded data can then be modified.

Radar too is a very useful means of measuring rainfall rates, and it is attractive in that data can be collected in considerable detail over a large area almost instantly. Although the accuracy of measuring rainfall rate at a single point by radar does not compare favourably with the best of rain gauges, radar may offer more representative statistics of rainfall rates over an area of several thousand square kilometres than is practicable using an array of rain gauges. A high-definition radar can give measurements of the most intense rain at the centre of a raincell where the possibility of that cell passing directly over a given rain gauge would be small. Furthermore, rainfall data collected over a large area for a short time may, to some extent, be a substitute for data collected at a point for a long time. When it comes to collecting information on rainfall rates over large areas at heights well above ground level, the radar is unchallenged.

Certain difficulties in the use of radar to measure rainfall rates accurately are as follows. First, although a short-wavelength (e.g. 3 cm) radar is more attractive in giving high angular resolution from a relatively small antenna, these wavelengths suffer considerably more *attenuation* than longer wavelengths (e.g. 10 cm). Indeed, the attenuation at 3 cm wavelengths is almost two orders of magnitude greater than the attenuation at 10 cm (see Fig. 3.12), and so attenuation by nearby raincells may cause the complete obscuration of radar returns from distant raincells. A wavelength of 5 cm is sometimes used as a compromise, but 10 cm is better, given a suitably large antenna to obtain adequate angular resolution.

Secondly, because different values of reflectivity factor can be associated with a given rainfall rate according to the *drop size distribution,* the calculated rainfall rate values may be in error by a factor of more than two (see Section 3.5). This error may be reduced by measuring the difference of the radar reflectivity in vertical and horizontal polarisations.[83] The differential reflectivity occurs because the raindrops fall as oblate spheroids with their axes of symmetry vertical (see Section 3.2). The larger drops are more oblate than the smaller drops, and so present a higher reflectivity for horizontally-polarised energy than for vertically-polarised energy. The differential reflectivity may be several decibels (see Fig. 3.16). By this means the drop size distribution is measured, rather than assumed, and the rainfall rate then calculated.

Thirdly, a conventional radar does not enable a *distinction between water drops and ice particles* to be made. Indeed the two may be present together, and wet ice (sleet) may also be present. Both melting snow and wet hail hold water in large drops and because the reflectivity factor is proportional to the sixth power of the drop size, the radar echoes are much larger than for the same total water content per unit volume in the form of rain. A sharp boundary between a region of rain and one containing large drops of wet snow or hail may be located using a dual-frequency radar system (e.g. at 10 cm and 3 cm together)[84]. In principle it is possible to measure the drop sizes by this method, but in practice the sensitivity of the technique is inadequate. Dual-polarisation radar has been used successfully to distinguish rain from ice particles.[52]

Very many *statistical distributions* of rainfall rates have been obtained throughout the world. These vary considerably from one location to another, and normally it would be preferable to use data collected locally, but, for places where these are not available, the CCIR have established five rain-statistics curves, and have identified general regions of the world to which they apply.[316] These are indicated in Fig. 3.8 and 3.9. The rainfall rates shown are the result of integration over one minute.

Data collected from rapid-response rain gauges[81] over 5 years (1970–1974 inclusive) at a site in the UK have been used to prepare statistics with *various integration times* from 10s to 1h.[59] The results, as shown in Fig. 3.10, may be used as a means of comparing cumulative distributions obtained from gauges with different integration times. The curve for the appropriate CCIR Climate 4 (which is for clock-minute measurement) is shown for comparison. From these curves it may be concluded that the rainfall rates that occur for a given time percentage p when measured with an integration time of 1 min occur for about $2p$ to $4p\%$ of time when measured with an integration time of 1h. Plotting the distribution curves separately for each of the 5 years covered in the study showed a range of about 35% above and below the mean rainfall rates for time percentages of 0·0003 to 0·03; and this was about the same whether using data integrated over 2 min or 10 s. A knowledge of this variability is of considerable interest to those wishing to use rainfall rate data for microwave communication planning since incorrect decisions may easily be made from observations conducted over only a short period of time.

For planning purposes there is a requirement to estimate the probability of attenuation due to precipitation exceeding certain levels in the *most unfavourable* ('*worst*') *month*, rather than in an average year. It is sometimes stated that the time percentage for which a given rainfall rate occurs in the worst month of an average year is four times greater than the time percentage for which it occurs in the whole year.

However, this factor is dependent on climate and on the rainfall rate selected for consideration and needs further clarification. Because the most intense rainstorms which so much disrupt communications

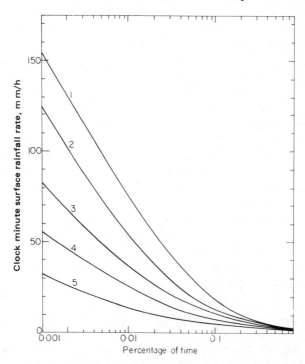

Fig. 3.8. *Percentage of an average year for which rainfall rate is exceeded for the five CCIR rain climates of Fig. 3.9* [Courtesy CCIR[316]]

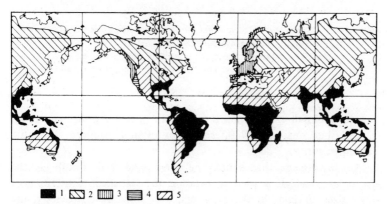

Fig. 3.9. *Regions corresponding to rainfall rate distributions of Fig. 3.8*
[Courtesy CCIR[316]]

are rare, it is preferable (where data allow) to consider rainfall rate statistics in terms of, for example, a 20-year period. Two factors may be distinguished; first, F_{20}, the ratio of the percentage p.a. of the worst month in a 20-year period for which a given rainfall rate is exceeded in that period to the percentage p.a. of the average year for which the same rate is exceeded; secondly, F_{av}, the ratio of the percentage of the worst month in an average year for which a given rainfall rate is exceeded to the percentage of an average year for which the same rate is exceeded. Fig. 3.11 shows the way in which these have been found to vary with rainfall rate in the UK using rapid response rain gauges.[85] For very large rainfall rates F_{20} becomes 240. This means that the rainfall occurs in a single time period which forms p_1% of that month or $p_1/240$% of 20 years. Similarly, F_{av} becomes 12, since the most unfavourable event in the average year would be p_2% of the month in which it occurs and $p_2/12$% of the year in which it occurs.

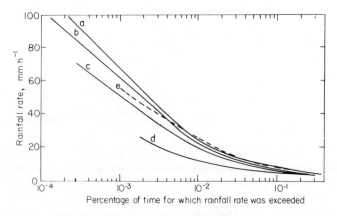

Fig. 3.10. *Distribution of rainfall rate for different integration times*
Based on UK data, averaged over 5 years
a 10 s   *b* 1 min   *c* 5 min   *d* 1 h   *e* CCIR climate 4 (see Fig. 3.8)
[Based on data from Harden, Norbury and White[59]]

For low rainfall rate events, the factors F_{20} and F_{av} become about 3 or 4, but such rainfall rates are not of much concern for radio communications.

An instrument that should be mentioned here briefly is the *distrometer*, which is used to measure the statistical distribution of drop sizes in rain. In the most common form of distrometer, the momentum of impact of the raindrops is measured as they strike a diaphragm.[86] The collecting area has to be small enough to ensure that only one raindrop arrives at a time, and the impact should be closely normal to the surface. A limitation to devices of the impact type is that the response to small drops is very poor. Because of the importance

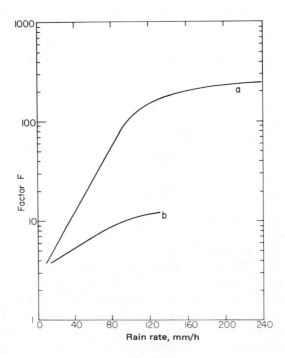

Fig. 3.11.   *Conversion factors for worst-month statistics*

$$a \quad F_{20} = \frac{\text{percentage of worst month}}{\text{percentage of average year}}$$

$$b \quad F_{av} = \frac{\text{percentage of worst month in an average year}}{\text{percentage of average year}}$$

(here the 'worst month' is that for which the rainfall rate of the abscissa is exceeded in a 20-year period)

[Based on data from Harden Norbury and White[85]]

of these small drops in their attenuation and scattering of millimetre waves, a more sensitive distrometer has been produced in which measurements are made of the interruption to a light beam caused by falling raindrops.[87] In yet another device the shadow of particles is observed on a large number of optical sensors.[88] This has been developed for observation of the shape as well as the sizes of particles and is used mounted in aircraft.

3.4 Attenuation

Attenuation due to *rain* arises from the absorption of the energy in the water droplets and from the scattering of energy out of the radio beam. In this Section the two effects are not separated, but in the next Section consideration is given to sideways and backwards scatter specifically.

The total attenuation A_R due to rainfall over a path of length r_0, and the specific attenuation due to rain $\gamma_R(r)$ (dB/km) along the path are related by

$$A_R = \int_0^{r_0} \gamma_R(r)\,dr \qquad \text{dB} \tag{3.5}$$

The attenuation may be evaluated approximately by using scattering theory and assuming the raindrops to be spherical.[89-91] This approach leads to expressions of the form[92]

$$\gamma_R = K_R R^a \qquad \text{dB/km} \tag{3.6}$$

where the specific attenuation coefficient K_R (dB/km/mm/h) and a depend on radio frequency, temperature and drop size distribution. Such relationships are plotted in Fig. 3.12 as a function of frequency.[93, 324] For all rainfall rates, the attenuation increases with frequency quite rapidly up to about 100 GHz. At about 200 GHz, the attenuation due to rain reaches a peak, and then shows a slight decline until, at optical frequencies, the values are almost constant. The curves have been obtained on the basis of experimental data on raindrop size distributions obtained by Laws and Parsons[94] and the terminal velocity of raindrops[79] using an empirical model for the refractive index of water at 20°C.[95] Temperature changes of 20°C may give departures from these curves of as much as 20% at frequencies below 2 GHz and at attenuations less than 0·1 dB/km. Suitable correction factors may be applied.[89]

Considerable study has been made of the validity of the experimental data and of the hypotheses used to calculate the attenuation due to rainfall, and of the applicability of these calculations to real rainfall situations.[90, 96-100] Measurements carried out at millimetre wavelengths in natural rain and in simulated rain are in satisfactory agreement with the theoretical calculations.[101] Many of the earlier measurements at longer wavelengths differ from theory mainly because of limitations in the rain-rate sampling techniques employed.[102]

The assumption made for Fig. 3.12 that raindrops are spherical is approximately true for small drops, but larger drops fall as oblate spheroids (see Section 3.2). One consequence of this is that *the attenuation for horizontally-polarised waves is greater than that for vertically-polarised waves.*[103, 104] The effect has been examined theoretically[104–106] and it has been shown that the ratio of the attenuations (dB) for the two polarisations is between 1·05 and 1·35 in the range 10 to 80 GHz for rainfall rates up to 200 mm/h.[85] This is entirely consistent with measurements made on line-of-sight paths at 13, 18 and 20 GHz.[107, 108]

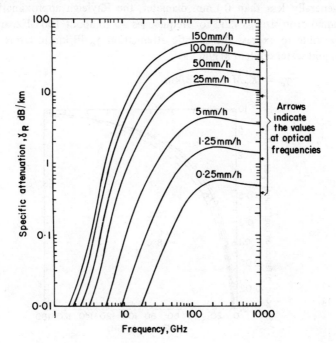

Fig. 3.12.   *Specific attenuation due to rain*

[Courtesy CCIR[324]]

Using a Laws and Parsons drop size distribution[94] gives a better explanation of the differential attenuation results than those put forward by Joss.[107, 109]

As an alternative to expressing the specific attenuation in terms of rainfall rate, as in eqn. 3.5, relationships have been suggested between γ_R and the *liquid water content* $M\,\text{g/m}^3$ of the rain of the general form

$$\gamma_R = aM^b \qquad \text{dB/km} \qquad (3.7)$$

The values of a and b are a function of drop size distribution, radio frequency and temperature. From examination of nearly 400 drop size samples, it has been suggested that $a = 0.36$ and $b = 1.27$ for a wavelength of 3·2 cm and temperature of 10°C,[110] but other results show a to lie between 0·006 and 16·3, and b to lie between 1·08 and 1·83 for the same wavelength and temperature but different rain conditions.[73] The tendency is to use relationships of the form of eqn. 3.6 rather than 3.7.

For *clouds or fog* consisting entirely of small *water* droplets, generally less than 0·1 mm diameter, the Rayleigh approximation of small drop sizes is valid for frequencies up to about 100 GHz, and it is possible to express the specific attenuation γ_c dB/km in terms of the liquid water content M g/m³ as[89]

$$\gamma_c = K_c M \qquad \text{dB/km} \qquad (3.8)$$

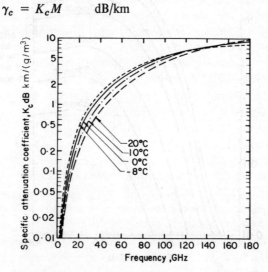

Fig. 3.13.   *Theoretical attenuation by water cloud at various temperatures as a function of frequency*

[Courtesy CCIR[324]. Based on Gunn and East[92]]

where K_c, dB/km/(g/m³), is the specific attenuation coefficient. Values of K_c are plotted as a function of frequency in Fig. 3.13 for various temperatures from $-8°C$ to $20°C$, based on theory.[92] Values of M for cloud vary between about 0·1 g/m³ and 5 g/m³, and values of M for different distances of visibility in fog are indicated in Fig. 3.7. Because of the much smaller complex permittivity, clouds consisting entirely of *ice particles* give attenuation about two orders of magnitude smaller than clouds of water drops of the same water content for frequencies up to 35 GHz.[89, 111] At higher frequencies the contribution of ice particle clouds to attenuation may be significant.

It has been reported that *dry snow* causes little attenuation for frequencies less than about 50 GHz, but that attenuation resulting from dry snow may exceed the attenuation produced by rain of equivalent water content at frequencies above about 60 GHz.[112] Although (melting) *wet snow* will have very large radar reflectivity due to the dependence on the sixth power of the particle diameters, the specific attenuation will still be approximately proportional to the mass of liquid water, and may be determined by eqn. 3.7. Attenuation by *hail* may be significant at frequencies as low as 2 GHz, although only at percentages of the time smaller than 0·001 in most climate regions.[321]

Such attenuation as is caused by clouds of *sand and dust* particles is due to scattering. The absorption is negligible. Measurements of simulated dust and sand conditions at 10 GHz have shown that the specific attenuation would be less than 0·1 dB/km for sand and 0·4 dB/km for clay dust where concentrations are less than 10^{-5} g/m^3.[113] In severe storms, the attenuation of microwave communication links might be rather more than this; but for ground-to-satellite links, where the effective path length through the dust storm may generally be expected to be less than 3 km, the attenuation of microwave signals is expected to be less than 1 dB.

At frequencies above about 50 GHz, the *attenuation due to atmospheric gases* exceeds the attenuation associated with even the most intense rainfall, but the gaseous attenuation on long radio paths through the atmosphere is significant once above about 15 GHz. The main atmospheric constituents producing attenuation are neutral oxygen and water vapour, although the presence of molecular complexes in water vapour may also have an effect at frequencies above about 100 GHz. The total gaseous absorption in the atmosphere A_g(dB) over a path of length r_0 (km) is

$$A_g = \int_0^{r_0} \{\gamma_0(r) + \gamma_w(r)\} \, dr \qquad \text{dB} \qquad (3.9)$$

where $\gamma_0(r)$ and $\gamma_w(r)$ are, respectively, the specific attenuation (dB/km) due to oxygen and water vapour under the conditions of pressure, temperature and humidity at a distance r along the path. The contribution from these two gases is indicated in Fig. 3.14 for the normally quoted conditions of pressure, temperature and humidity. Water vapour absorption has resonant peaks at frequencies of 22·5, 183 and 320 GHz, and oxygen absorption has a broad peak at 60 GHz and a narrow one at 119 GHz. All these peaks are pressure broadened in the lower atmosphere, and that at 60 GHz is particularly prominent being the composite of a number of lines.

The absorption due to oxygen at any given height, like the proportion of oxygen in the atmosphere, is approximately constant with time. However, the water-vapour content is very variable, and so too is the associated attenuation. The average value of water vapour concentration

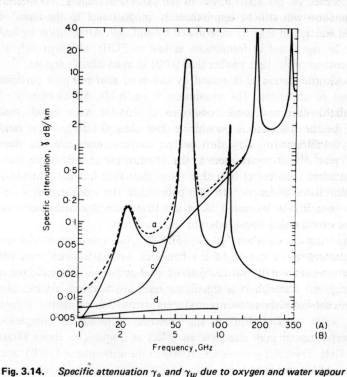

Fig. 3.14. *Specific attenuation γ_0 and γ_w due to oxygen and water vapour*

$a\ \gamma_0 + \gamma_w$ for $f > 10\,\text{GHz}$ ⎫

$b\ \gamma_w$ for $f > 10\,\text{GHz}$ ⎬ scale A

$c\ \gamma_0$ for $f > 10\,\text{GHz}$ ⎭

$d\ \gamma_0$ for $f < 10\,\text{GHz}$ · scale B

Pressure: 1 atm (1013 mb)

Temperature: 20°C

Water vapour density: 7·5 gm/m³ (i.e. $\gamma_w = \gamma_{w7\cdot5}$)

[Based on CCIR[322]]

at ground level for temperate climates is about $7\cdot5\,\text{g/m}^3$, but the amount of water vapour that the air can hold is heavily dependent on the air temperature (see Fig. 2.1). As a first approximation, the specific attenuation may be considered as proportional to the atmospheric water vapour concentration m (g/m³), i.e. $\gamma_w = (m/7\cdot5)\gamma_{w7\cdot5}$ where

$\gamma_{w7.5}$ is the value obtained from Fig. 3.14. For a radio path close to the ground, pressure and temperature variations along the path have negligible effect on the gaseous absorption, although allowance must sometimes be made for humidity variations over a very long path above varied types of ground and above sea or lakes. For an earth-space path, the variation of atmospheric pressure, temperature and humidity with height must be taken into account. This will be considered in Section 5.3.1.

3.5 Scatter

In addition to attenuation, hydrometeors can cause scattering of energy into or out of a radio beam which in the sideways direction may cause very troublesome interference between radio paths, and in the backwards direction may be measured by radar. The interference problem is particularly acute between ground-to-ground (terrestrial) links and earth-to-satellite links (see Section 8.3.3). The main cause of this scattering is rain, but scatter from clouds of ice particles and from wet snow or wet hail is also important.

The general equation for the radio power scattered in the volume of atmosphere common to the beam of transmitting and receiving antennae, is given by eqns. 2.46 and 2.47. For a *single spherical drop of water*, with diameter D(mm) small compared with the radio wavelength λ(m), the Rayleigh scattering cross-section σ(m^2) is given by

$$\sigma = \frac{\pi^5 |K|^2}{\lambda^4} \times 10^{-18} \sin^2 (\beta) D^6 \qquad (3.10)$$

where $|K| = |(n^2 - 1)/(n^2 + 1)|$, n^2 is the complex permittivity of water (see Section 4.2.3) and β is the angle between the direction of scattering and the electric field in the incident wave. At SHF, $|K|^2$ may be taken to be 0·93 and 0·20 for water and ice particles, respectively, so that scatter from water drops will produce nearly five times the scattering cross-section of ice particles of the same size and number per unit volume. For the practical case of rain, the drops have a range of different sizes in a statistical distribution for which $N(D)$ is the number of drops of diameter D per unit interval of diameter per unit volume (see eqn. 3.1). Consequently, the *scattering cross-section per unit volume*, η(m^2/m^3) is given by

$$\eta = \frac{\pi^5 |K|^2}{\lambda^4} \sin^2 \beta \times 10^{-18} \int N(D) D^6 dD \qquad (3.11)$$

where integration is over all drop sizes, and where the term

$$Z = \int N(D)D^6 dD \qquad \text{mm}^6/\text{m}^3 \qquad (3.12)$$

is generally known as the reflectivity factor. Again small spherical drops are assumed.

In the rather special case of radar backscatter (as when used to measure rainfall rate), $\beta = \pi/2$. The volume V in eqn. 2.47 is then the radar pulse volume, which is normally assumed to be uniformly filled with rain of reflectivity factor Z. Clearly, for long wavelength radars with small antennae, the rain will not usually be uniform in the large pulse volume and some average will be obtained. In the case of horizontally-propagated vertically-polarised waves, scatter is omnidirectional in the horizontal plane (i.e. $\beta = \pi/2$); but if the deviation of the scattered wave from the horizontal plane is large, then the influence of the $\sin^2\beta$ term in eqn. 3.11 must be taken into account. Similarly, horizontally-polarised waves will not be omnidirectionally scattered in the horizontal plane.

Strictly, eqns. 3.11 and 3.12 apply only for small drops, but, for frequencies greater than about 5 GHz, Mie scatter theory must be applied in place of Rayleigh scatter theory because the drops cannot then be considered small compared with the wavelength. The ratio of the calculated reflectivity factors based on the two theories is indicated in Fig. 3.15 as a function of rainfall rate[114] assuming a Marshall and Palmer drop size distribution (see Section 3.2). Due account has been taken of the variation of refractive index with frequency. The Figure shows how the two theories diverge for higher rainfall rates (and for higher frequencies) as the larger drops in the drop size distribution cease to be small compared with the wavelength. Furthermore, although the smaller drops tend to be nearly spherical as they fall, the larger drops are oblate (see Section 3.2) which causes the reflectivity factor for a horizontally-polarised wave Z_H to be greater than that for a vertically-polarised wave Z_V. The ratio of the two, known as the differential reflectivity factor Z_{DR} is shown in Fig. 3.16 as a function of the diameter D_0 of the median-volume drop of the drop size distribution.[52] For D_0 greater than about 3 mm, the form of this curve is very dependent on the maximum drop size. It also depends on the radio wavelength since resonance within the drop becomes significant for large drops.

In the absence of detailed statistics of the spatial distribution of the reflectivity factor Z, reliance has been placed on such information as

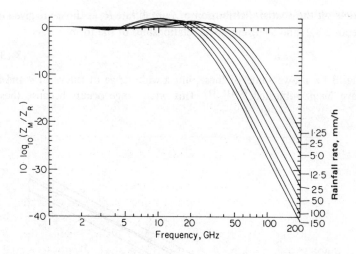

Fig. 3.15. *Ratio of reflectivity factors calculated by Mie scatter theory Z_M and Rayleigh scatter theory Z_R*
[Based on data, from Devasirvatham and Hodge[114]]

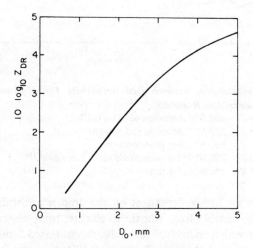

Fig. 3.16. *Variation of differential reflectivity factor Z_{DR} with median volume drop diameter D_0 for equivalent sphere*
Assuming a maximum drop diameter of 9 cm and a wavelength of 10 cm

exists on the spatial distribution of rainfall rate R, as those are given in Section 3.2, and using empirical relationships of the form

$$Z = aR^b \tag{3.13}$$

Fig. 3.17 shows some of these, but a wide range of values of a and b have been published.[72, 115-117] This wide range occurs because these

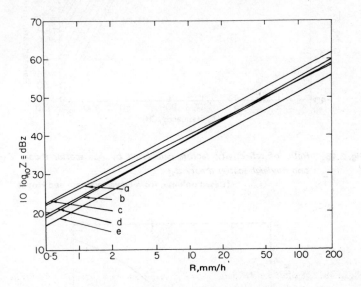

Fig. 3.17.   *Relationships between radar reflectivity factor $Z(mm^6/m^3)$ and rainfall rate $R\ (mm/h)$*
a $Z = 400\ R^{1\cdot4}$, recommended by CCIR[316]
b $Z = 220\ R^{1\cdot6}$, Marshall and Palmer[66]
c $Z = 500\ R^{1\cdot5}$, for thunderstorms ⎫
d $Z = 250\ R^{1\cdot5}$, for widespread rain ⎬ *Joss et al.[116]*
e $Z = 140\ R^{1\cdot5}$, for drizzle ⎭

relationships are highly dependent on the drop size distribution, and this is highly variable from climate to climate, from one rain type to another and within individual rain cells. In Section 3.3 mention was made of using relationships such as eqn. 3.13 to compute rainfall rates from reflectivity, but this may lead to error factors more than two due to the variability of drop size distribution from that assumed for the computation.

Use has also been made of relationships of the form of eqn. 3.13 to estimate interference due to scattering from rain based on rainfall statistics (see Section 8.3.3). However, some directly-measured data do exist on the way in which the reflectivity factor varies with height for different time percentages, and this has been used to form models for all climates.[316] Fig. 3.18 shows results from three studies, and in each case the curves show the Z values exceeded for 0·01% of time.

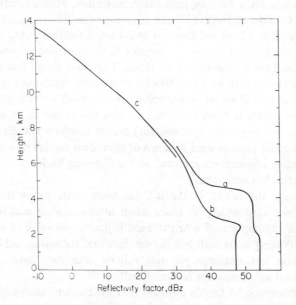

Fig. 3.18. *Experimental data showing variation with height of reflectivity factor exceeded for 0·01% of time*
a Data from Japan[118]
b Data from UK[119]
c Data from USA/Canada[121]

Curve *a* is deduced from a 5 GHz radar study in Japan, using a fixed 0·6° wide beam elevated at 10° from horizontal.[118] The shape of the curve is taken from values of Z exceeded at certain heights for 0·01% of time, and the value at ground has been displaced to fit the surface rainfall rate data for climate 2 (as given in Fig. 3.8) using eqn. 3.13. The effect of the melting layer 'bright band' at about 4 km is prominent.

Curve *b* is taken from a 3 GHz radar study in the south of England.[119] Raincells were scanned by the 0·25° radar beam at a succession of elevation angles, and a series of $Z(h)$ curves plotted for different surface reflectivity factor values. Curve *b* was selected from these using eqn. 3.13 and the rain (climate 4) statistics of Fig. 3.8 to determine the value of Z at ground. The melting layer is again a prominent feature, this time at 2·5 km height. From a similar 3 GHz radar study in coastal

Virginia,[120] the results were expressed in terms of the peak values of reflectivity factor in raincells and so comparison with the curves of Fig. 3.18 may be misleading.

Curve *c* was obtained from forward-scatter beam-swinging measurements made on a 500 km path between Boston, Massachusetts, and Ottawa, Ontario, at 16 GHz.[121] The common volume of the scatter path had a length of 12 km and diameter of 0·7 km at the centre, but curve *c* assumes the effective horizontal distance of the scattering volume to be limited to 3 km by the size of ice cloud. The curve shows that scatter from ice clouds may be a problem for some communications systems. Computation of Z values on the assumption of small water drops, even when the scatter is from ice cloud, enables the values to be used in eqn. 3.13 for any rain (hydrometeor) scatter computation. Similarly, the data could be expressed in terms of equivalent rainfall rate, so long as the same conventions are used in transforming back to scattering cross-section per unit volume.

Although the height of the $0°C$ isotherm varies during the year, it does not vary widely at times when intense scatter occurs from rain. The prominence of a bright band is due to the value of $|K|^2$ in eqn. 3.10 being some 6 dB less for ice than that for water, and due to there being less raindrops per unit volume once they have reached terminal velocity below the region of melting.[115]

The discussion so far has assumed that the incident energy is plane-polarised. If, on the other hand, the incident energy were circularly-polarised, then the energy scattered directly forward would retain the same polarisation, whereas energy scattered directly backwards would experience a reversal in the direction of polarisation. However, for reflection from a large conducting surface, the sense of polarisation is maintained. This property may be exploited in conventional radar use as a rain-echo-cancellation technique, since returns from conducting surfaces are accepted by the radar antenna, whereas those from rain are not. Circularly-polarised waves scattered by rain through a right angle would become linearly-polarised normal to the scattering plane.

3.6 Thermal noise emission

Any absorbing medium at a temperature above absolute zero is also a source of thermal noise power radiation. For a body which absorbs all radiations falling on it, a 'black body', the radiated power $P(W)$ at temperature $T(K)$ is given by

$$P = kTB \tag{3.14}$$

where $k = 1.38 \times 10^{-23}$ W/K/Hz is Boltzman's constant, and B(Hz) is the bandwidth. For a partially-absorbing medium at uniform temperature T_a having a loss factor L (i.e. an absorption coefficient α and transmission coefficient $\beta = 1/L$), the effective (measurable) noise temperature (or equivalent black-body temperature) T_b is given by

$$T_b = \left(1 - \frac{1}{L}\right) T_a + \frac{1}{L} T_0 \tag{3.15}$$

where T_0 is the temperature behind the absorbing medium. For an opaque medium $T_b = T_a$, $\alpha = 1$, $\beta = 0$, $L = \infty$ and no noise power is transmitted through the medium. By contrast, for a transparent medium $T_b = T_0$, $\alpha = 0$, $\beta = 1$, $L = 1$ and the medium contributes no noise power.

By way of illustration, the effective noise temperature of the atmosphere due to absorption by clouds or precipitation is the same as that of the medium itself T_a (i.e. about 270 K) when the attenuation is large, but falls to the value of T_0 (i.e. about 4·7 K) at zenith in the absence of any attenuation, e.g. below about 10 GHz.

Eqn. 3.15 is the basis for using *radiometers* (sensitive radio receivers) to measure the atmospheric noise power in order to deduce the *atmospheric attenuation* A_a(dB) along a slant path from ground (see Section 5.3).

Either the power itself may be determined or T_b may be compared with a reference temperature source, since the ratio of the temperatures will be equal to the ratio of the powers received. If a power-law detector is used in the radiometer, then the output voltage will be proportional to noise temperature, as shown in Fig. 3.19. The attenuation A_a (where $A_a = 10 \log_{10} L$) may then be obtained by rearranging eqn. 3.15 as

$$L = \frac{1}{\beta} = \frac{T_a - T_0}{T_a - T_b} = \frac{V_{max} - V_{min}}{V_{max} - V} \tag{3.16}$$

If there is a loss factor L_g owing to attenuation in the radiometer antenna and feed system (at temperature T_g) which is not in the feed line of the reference temperature souce, the measured effective atmospheric noise temperature T_e will become

$$T_e = T_g \left(1 - \frac{1}{L_g}\right) + \frac{1}{L_g} T_b \tag{3.17}$$

The effective noise power from the atmosphere may have to be taken into account on an earth-space path, see Section 8.2. To assess the noise power using attenuation statistics, or to interpret data from radiometers, values of T_a and T_0 have to be assumed. Normally, for rain calculations

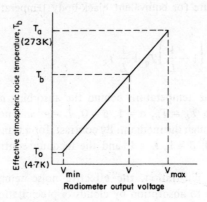

Fig. 3.19. *Radiometer characteristic*

T_a is taken as 273 K (sometimes as high as 290 K), and T_0 as 4·7 K (sometimes approximated as zero), although it is preferable to calculate the rain temperature from the ground-level temperature and its assumed change with height. Where the attenuation is due to high-level cloud alone, the temperature T_a will be much lower than 270 K. In the estimation of antenna temperatures, radiation (reflected or direct) from the surface of the earth into antenna side lobes (or even the main lobe) must also be considered, as well as extra-terrestrial radiation. Normally, care is taken to ensure that the radiometer beam is well above the local horizon.

In the absence of rain or clouds, the effective atmospheric noise temperature at an elevation angle θ for a horizontally-stratified atmosphere is given by

$$T_b(\theta) = T_a(1 - \beta_0 \operatorname{cosec} \theta) \tag{3.18}$$

for elevation angles greater than 15°, where β_0 is the transmission coefficient of the atmosphere in the zenith direction and where $T_a \gg T_0$. Fig. 3.20 shows the atmospheric noise temperature for an infinitely narrow beam at 0°, 10° and 90° elevation, and for frequencies between 1 and 100 GHz. The curves are based on a theoretical model, for a pressure of 1013 mb, a temperature of 20°C and water vapour concentrations of 3, 10 and 17 g/m³ at ground level (corresponding to relative humidities of 17, 58 and 98%, respectively), assuming decreases with height of water vapour concentration, oxygen pressure and temperature applicable to middle lattitudes. Allowance has been made

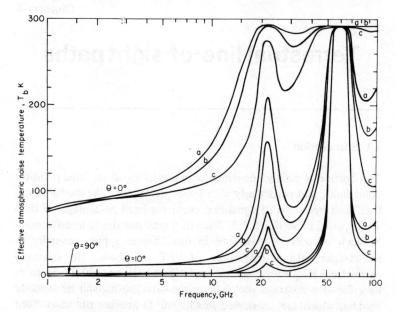

Fig. 3.20.   *Effective atmospheric noise temperature in the absence of cloud or rain as function of frequency for a narrow antenna beam for elevation angles of 0°, 10° and 90° and for (a) 17, (b) 10, and (c) 3 g/m³ and 1013 mb and 20°C at ground level*

[Based on CCIR³²³]

for refraction at low elevation angles. The maximum change in emission noise due to changes in assumed ground-level temperature (with a constant 10 g/m³ at ground level) is ± 3·5% for the range 13 to 27°C (within the 1 to 100 GHz range of Fig. 3.20). The curves are in agreement with the limited amount of experimental data available. By these means the atmospheric noise temperature and slant path attenuation may be obtained at various elevation angles for a given locality taking into account its climatic characteristics.

Chapter 4

Terrestrial line-of-sight paths

4.1 Introduction

This Section is mainly concerned with loss of signal on fixed point-to-point links, and particularly with losses greater than the median level. Small enhancements of signal that occur for small percentages of time are discussed in Section 8.3.2., since they may give rise to interference.

Much of what is described in this Chapter applies generally to line-of-sight links at frequencies of VHF and above. The relatively narrowband frequency channels available at VHF and UHF tend to be used for area coverage applications, e.g. broadcasting and land-mobile services, which are considered in Chapter 7, whereas the main fixed point-to-point services employ frequencies above UHF where relatively wideband multi-channel services can be provided. This emphasis is reflected in the content of the present Chapter.

When the two terminals of a terrestrial path are within radio line-of-sight, several influences may cause loss of signal. If the line-of-sight of the system is close to the ground, large-size obstacles or hills, then diffraction losses may become important even though the direct line-of-sight is not obscured. With any changes of effective-earth-radius factor, due to refraction, the path may then be subject to diffraction fading. Criteria for avoiding this are considered in Section 4.2.1. When the line-of-sight is well above the earth's surface, so that diffraction losses are avoided, fading may occur due to interference between the direct line-of-sight component of field and that reflected from the ground, atmospheric layers or buildings (see Section 4.2.3 and 4.3). Measurements have shown that as many as six multipath components may exist at any one time,[122] but the most severe fading occurs when there are only two effective components of field strength with similar magnitude. Together with scintillation, multipath effects may give rise to short-term fading, although the latter are unlikely to be of consequence at frequencies below 10 GHz. At frequencies above UHF, attenuation due to absorption by oxygen and water vapour, and due to absorption and scattering by hydrometeors (rain, hail, snow and fog) may be even more important (see Section 4.4). Hydrometeor effects may give rise to longer-term signal attenuation.

The relative importance of fading due to rain and that due to multi-path effects depends on frequency, climate and path length. In general it can be said that multipath fading is the main factor causing attenuation below 10 GHz, whereas heavy rain is the main influence above this frequency, particularly on shorter links. Because in most climates multipath propagation normally occurs when there is no heavy rainfall, it is usually reasonable to add together the time percentages for which the two causes produce fades of a certain level.

4.2 Ground reflection losses

4.2.1 *Clearance of line-of-sight from terrain features*
If the terminals of a line-of-sight path are so low that parts of the path pass close to the surface of the earth, the transmission loss due to diffraction effects will be well in excess of the free-space value even if the path is not directly obstructed. An objective in planning terrestrial links is to elevate the terminals sufficiently to avoid these losses. The extent of the clearance of terrain features from the direct line-of-sight may be assessed in terms of *Fresnel zone ellipsoids* drawn around the path terminals (see Section 7.2).

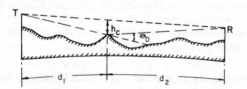

Fig. 4.1.   *Clearance height h_e of line-of-sight from terrain obstruction*
Path profile drawn for straight-ray model

Fig. 4.1 illustrates a line-of-sight radio path of length d which has a terrain feature distance d_1 from one terminal and d_2 from the other. The clearance height of the line-of-sight from the terrain feature is h_c when normal refraction is assumed. This height may be expressed in terms of the radius r_n of the nth Fresnel zone at that point, where

$$r_n = [n\lambda d_1 d_2 / d]^{1/2} \tag{4.1}$$

i.e.

$$h_c = \sqrt{n}\, r_1 \tag{4.2}$$

It is assumed here that the terrain feature is in the far field of the link antennae. If $d_1 = d_2 = d/2$ and $n = 1$, then the first Fresnel zone may be shown to have a maximum radius $r_{1\,max}$ of

$$r_{1\,max} = \tfrac{1}{2}(d\lambda)^{1/2} \tag{4.3}$$

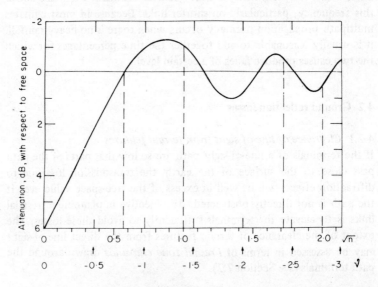

Fig. 4.2.   *Attenuation due to knife-edge diffraction as a function of the normalised clearance height $h_c/r_1 = \sqrt{n}$*
This function, when expressed in terms of v, is known as the Fresnel-Kirchhoff function, where $v = -\sqrt{2n}$

Fig. 4.2 illustrates the *diffraction loss* (attenuation with respect to free space) as a function of the clearance height h_c from a line-of-sight radio path normalised in terms of the first Fresnel zone radius r_1, i.e. $\sqrt{n} = h_c/r_1$. The Fresnel-Kirchhoff relationship shown in the Figure applies only for knife-edge diffraction (with no reflection from the edge of the screen). In place of the normalised clearance height $\sqrt{n}$ the diffraction loss is sometimes plotted as a function of a diffraction parameter $v = -\sqrt{2n}$, so that

$$v = -h_c(2d/\lambda d_1 d_2)^{1/2} \tag{4.4}$$

where

$$h_c = (\sin \theta_D)(d_1 d_2/d) \tag{4.5}$$

and θ_D is the angle indicated in Fig. 4.1. This point will be taken up again in Section 7.2 when considering paths obstructed by diffracting edges.

Fig. 4.2 shows that ensuring that 55% (usually taken as 60%) of the first Fresnel zone around a radio path is completely free of obstruction avoids a 6 dB diffraction loss compared with the grazing condition, $h_c = 0$. This 6 dB loss occurs for grazing incidence because half of the wave front, and so half the field, that would be present in the absence of the diffracting obstacle is obstructed. Achieving any additional clearance gives little further benefit. However, the atmospheric refraction on a line-of-sight path is time varying, and any temporary subrefraction will have the same effect as increasing the height of potentially-obstructing obstacles. Consequently, it is normal practice to ensure that 60% clearance occurs for an effective-earth-radius factor k less than unity to minimise (subrefractive) *diffraction fading*. Usually a k factor of 0·7 (i.e. 4500 km radius) is adopted. Analysis of 300 000 h of chart records from 21 radio-relay links in the UK planned more or less in accordance with this criterion has shown no instances of such fading.[314] A less conservative criterion, namely that all the first Fresnel zone should be unobscured for a k factor of 4/3, lead to subrefractive fading in Germany on paths longer than 100 km.[314] Certainly there are climates from which the statistical variation of the effective-earth-radius factor is very large, and account should be taken of such information in order to ensure link reliability. Clearly, from eqn. 4.1 a hill near the centre of a path is more likely to cause diffraction fading than a similar hill near a path terminal.

When allowing for the variation of refractive index gradient on radio paths it is desirable to plan for a *value of k averaged along a path* rather than the value of k at any particular point. The variability of this path-averaged value is less than the variability at a point. The path-averaged value of k which is exceeded for 99·9% of time is shown in Fig. 4.3 as a function of path length for continental temperate climates. Clearly, the longer the radio path, the greater is the minimum path-averaged value of k.[123] In hot and wet climates the values of k will be higher and for a dry climate they will be lower. For a path length of 50 km, the average values of k commonly taken as the lower limit for calculations are $k = 1$ for wet climates, $k = 0.8$ for temperate climates and $k = 0.6$ for desert climates.[124]

In planning a new radio path, the vertical section of the first Fresnel zone around the line-of-sight may be *plotted on a path profile*, using eqn. 4.1, as illustrated in Figs. 4.4a and b, for either a flat earth or curved 4/3-earth-radius representation. Fig. 4.4c shows a composite of Figs. 4.4a and b in which a straight-ray representation is used and the effective-earth-radius factor is allowed for only at points likely to

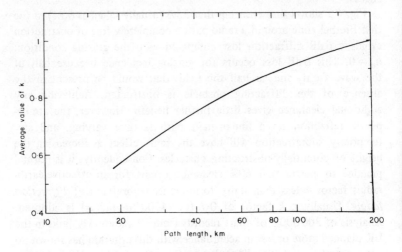

Fig. 4.3. *Path-averaged value of k exceeded for approximately 99·9% of the time (Continental temperature climate)*

[Courtesy CCIR[321]]

cause obstruction. In each case the first Fresnel zone is plotted for wavelengths of 0·1 m and 1 m for median refraction, $k = 4/3$, and Fig. 4.4c also shows the obscuration for $k = 2/3$. With $k = 4/3$ there is no obscuration for the 0·1 m wavelength, but there is obscuration for 1 m wavelength. With $k = 2/3$ there would be obscuration even at much shorter wavelengths, since the direct line-of-sight itself is almost obstructed at a distance of 55 km. To avoid this obscuration, the path terminals would have to be raised a further 30 m for operation at 0·1 m wavelength and 120 m for operation at 1 m wavelength. Information from maps is normally sufficient for plotting a path profile, but it may be necessary to carry out some form of local survey to complete critical sections satisfactorily. For any terrain between the path terminals covered with extensive woodland, it is appropriate to add the height of the trees to the ground height when drawing the path profile. Although attenuation through thin belts of trees is not great, it is severe if long distances of woodland are traversed (see Section 7.3.2). At VHF the tree tops normally present a high reflection coefficient. At frequencies above UHF, the presence of individual buildings will normally be significant.

4.2.2 Specular and diffuse reflection

On a line-of-sight path, where the terminal heights are above the region

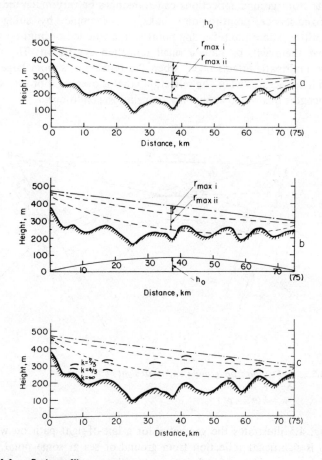

Fig. 4.4. *Path profile*
   a for flat-earth representation for $k = 4/3$
   b for straight-ray representation for $k = 4/3$
   c composite of *a* and *b* for $k = \infty$, $k = 4/3$ and $k = 2/3$
   In each case
   —.—.— represents the line-of-sight (direct way)
   - - - - - - represents the first Fresnel zone
        (i) for $d = 1\,\text{m}$, $f = 300\,\text{MHz}$
        (ii) for $d = 0.1\,\text{m}$, $f = 3000\,\text{MHz}$
   $h_0 = d^2/8a_e$, for path length d and effective earth radius a_e

where diffraction fields are dominant, the effect of ground or obstacles near to the path may give rise to reflected or scattered energy arriving at the receiver in addition to the direct ray. This may cause slight signal enhancement or severe fading as refractive index variations on the path change the relative phases of the different ray components. Multipath effects from ground reflections can sometimes be very much reduced on fixed-service point-to-point links, for example, by siting the transmitter where the reflecting point is in a wooded or built-up area, so that there will be only a small reflection coefficient. The most severe multipath effects occur when the reflection is from large flat smooth surfaces, as when the reflection point is over sea. For very short wavelengths this would apply only for calm sea conditions.

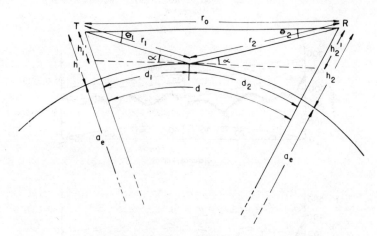

Fig. 4.5.   *Reflection from ground on line-of-sight path*
$$\Delta_r = r_1 + r_2 - r_0$$
$$\alpha = \tfrac{1}{2}(\theta_1 + \theta_2)$$

Fig. 4.5 illustrates the geometry for a line-of-sight path on which there is significant reflection from ground or sea at some point. For small grazing angles, the reflected wave will generally have some phase delay ϕ_Δ due to the path delay Δ_r, such that

$$\phi_\Delta = 2\pi\Delta_r/\lambda = 4\pi h_1' h_2'/\lambda d \tag{4.6}$$

and some phase delay ϕ_R associated with the *reflection coefficient* $R = |R|e^{j\phi_R}$ of the surface. The resultant field E for the direct and reflected rays is then

$$E = E_d(1 + Re^{j\phi_\Delta}) = E_d(1 + |R|e^{j(\phi_\Delta + \phi_r)}) \tag{4.7}$$

where E_d is the direct ray. The magnitude of the reflection coefficient $|R|$ and its phase ϕ_R will depend on the nature of the reflecting surface, the angle between the surface and the incident wave α (normally known as the grazing angle), the radio wavelength λ, any curvature of the surface and the degree of surface roughness.

In general, the field due to scattering from a moderately rough surface is the sum of a *specular component* and a *diffuse component*. For the former, the scattered energy is contained within a cone close to the direction for which the angle of reflection is equal to the angle of incidence, and its phase is coherent, i.e. its mean value can be determined for any point in space. This specular component is the result of the re-radiation of energy from points on a Fresnel ellipsoid which give rise to equal phase at the receiver. Such points will lie within a limited area on the path. Except in the special case of a perfectly smooth surface, the amplitude and phase exhibit small fluctuations in space, and also in time if there are any time variations of the surface shape (as in the case of the sea or wind-blown trees) or of the atmosphere above it. When a receiving antenna is moved in space, the relative phase of the direct and reflected wave will produce a regular interference pattern. Conversely, the diffuse component has little directivity and originates over a larger area of the scattering surface than that which produces the specular component. Its mean value is much less than that of the specular component, but its fluctuations are Rayleigh-distributed, and so have a large amplitude. Its phase is incoherent, i.e. at any point in space it may assume any value with equal probability. When a receiving antenna is moved in space, the received field will vary in a random manner and may be described by its statistical distribution and its correlation distance.

If the surface is very rough, there will be no specular component and the diffuse component will be uniformally distributed with angle. A moderately rough surface exhibits a *'glistening' area*, and the scattered field may be regarded as the sum of many components, each specularly reflected from facets of the rough surface. An optical analogy is that of moonlight reflected from a randomly-rippled water surface, although the extent of a Fresnel zone contributing to a reflecting facet is then much smaller than that required in the radio case. Situations have been considered in some detail for which

(*a*) both terminals are near ground level on a line-of-sight path and
(*b*) one terminal is near the ground surface and the other is very far away, e.g. a satellite on aircraft.[125]

For the first of the two cases, then if terminals have wide-beam antennae and are essentially symmetrically placed on the path, it may be shown that, if the grazing angle for specularly-reflected ray components is much less than the maximum slope of an elementary facet (as is normally the case for terrain or the sea surface), the scattered energy originates principally from the two regions in the neighbourhood of the link terminals, and that the mean-square of the reflection coefficient (i.e. the mean of the power reflected) is typically about half that for a smooth surface. If, however, the grazing angle is much larger than this, then the scattered energy originates from a zone surrounding the point of reflection given by geometrical optics, and the mean-square reflection coefficient may be close to unity. Whether or not the maximum slopes of the elementary facets are large compared to the grazing angle, the effective area over which reflection can occur, and the consequent reflected energy may, in practice, be limited by the narrowness of the antenna beams rather than the glistening area.

For a randomly-varying Gaussian-distributed moderately rough surface, the RMS specular term (for the specular direction) of the reflection coefficient is the product of that for a smooth earth and a *roughness factor* $f(\sigma_i)$ such that [125]

$$f(\sigma_i) = \exp -\tfrac{1}{2}[4\pi\sigma_i(\sin \alpha)/\lambda]^2 \qquad (4.8)$$

where σ_i is the standard deviation of the surface irregularities. For a very rough surface σ_i is large, and so $f(\sigma_i)$ tends to zero and only the diffuse contribution remains. Conversely, for a smooth surface σ_i is zero and $f(\sigma_i) = 1$. Moreover, there will then be no variations of amplitude or phase. A well established test for the surface to be considered smooth is the *Rayleigh criterion* [125]

$$H = 7 \cdot 2\lambda/\alpha \qquad (4.9)$$

where α is expressed in degrees, and H is the height between the top and bottom of the surface irregularities. For this condition there is a range of $\pi/2$ phase difference for energy scattered from these irregularities and so there is no direct cancellation of any contributions. However, assuming $H = 2\sigma_i$, then eqn. 4.8 gives $f(\sigma_i) = 0 \cdot 3$, which is rather small (i.e. the power reflected is 11 dB below that for a smooth surface). Rather more strict criteria, e.g. a phase difference of $\pi/4$ or $\pi/8$, giving factors of $3 \cdot 6$ or $1 \cdot 8$ in place of $7 \cdot 2$ in eqn. 4.9, and $f(\sigma_i)$ of $0 \cdot 73$ and $0 \cdot 93$ ($-2 \cdot 7$ dB and $-0 \cdot 7$ dB), respectively, have been considered to be more realistic. [126-128] Any of these criteria can only be a general guide, since true specular reflection applies only when H and α tend towards zero.

It is stated by some authors that ground reflections do not occur over land paths because of terrain roughness. Although this may normally be true above UHF, it depends on the wavelength and the angle of incidence. This is indicated in Table 4.1. From this Table some

Table 4.1 *Rayleigh criterion of surface roughness values of H, from eqn. 4.9*

	$f =$ $\lambda =$	100 MHz 3 m	1000 MHz 30 cm	10 000 MHz 3 cm
For $\alpha = 1°$		20 m	2 m	20 cm
For $\alpha = 0 \cdot 1°$		200 m	20 m	2 m

indication may be gained of the relative importance of various terrain features, e.g. buildings and trees, as a function of wavelength and grazing angle. It should be recalled that for a flat earth $\alpha \simeq (h_1 + h_2)/d$. For open sea conditions (remote from land) probable maximum wave heights are indicated as a function of wind speed in Table 4.2.

Table 4.2 *Wave heights in open sea and wind force*

Wind force (Beaufort)	Mean wind speed	Description	Probable maximum wave height
	km/h		m
0	—	calm	—
1	3·7	light air	0·1
2	9·3	light breeze	0·3
3	17	gentle breeze	1·0
4	26	moderate breeze	1·5
5	35	fresh breeze	2·5
6	46	strong breeze	4·0
7	57	near gale	6·0
8	69	gale	7·5
9	81	strong gale	10·0
10	100	storm	12·5
11	110	violent storm	16·0
12	—	hurricane	—

Since eqn. 4.8 is more useful than the Rayleigh criterion, this is evaluated for some spot values of σ_i/λ and $\alpha°$ in Table 4.3. Evidently for $\alpha < 1°$, the specular contribution remains within 2 dB of the smooth earth conditions so long as $\sigma_i < 3\lambda$.

Table 4.3   *Effect of surface roughness on specular reflection*

$\sigma_i/\lambda =$	1	3	10	30
For $\alpha = 1°$	$-0\cdot2$	-2	-20	-200
For $\alpha = 0\cdot1°$	$-0\cdot002$	$-0\cdot02$	$-0\cdot2$	-2

Values of $f(\sigma_i)$, in decibels, from eqn. 4.8

4.2.3 *Reflection from a smooth earth*

For an assumed smooth earth the reflection coefficient R is that of a plane surface of infinite extent (known as the Fresnel reflection coefficient) ρ, multiplied by a factor $f(\nu)$, for the finite extent of the reflecting surface and a divergence factor D, for any curvature of the reflecting surface. The Fresnel reflection coefficient is given by

$$\rho_V = \frac{n^2 \sin \alpha - (n^2 - \cos^2\alpha)^{1/2}}{n^2 \sin \alpha + (n^2 - \cos^2\alpha)^{1/2}} \tag{4.10}$$

for vertical polarisation, and

$$\rho_H = \frac{\sin \alpha - (n^2 - \cos^2\alpha)^{1/2}}{\sin \alpha + (n^2 - \cos^2\alpha)^{1/2}} \tag{4.11}$$

for horizontal polarisation, where α is the grazing angle and n is the refractive index. In practice, the value of ρ is highly dependent on the *permittivity* (or dielectric constant) relative to free space ϵ_r, and the *conductivity* σ S/m of the ground, since the complex permittivity $n^2 = \epsilon_r - j\sigma/\omega\epsilon_0 = \epsilon_r - j\,60\sigma\lambda$. Here ϵ_0 is the permittivity of free space. The relative permeability is assumed to be unity. Fig. 4.6 shows how very markedly the permittivity and conductivity change as functions of frequency in the region from 10 MHz to 100 GHz.

Because n^2 is complex, unless the ground can be assumed to be a perfect dielectric as is almost true for dry desert conditions, so too are ρ and R, i.e. $\rho = |\rho|e^{j\phi_\rho}$ (here $\phi_\rho = \phi_R$, since the factors for surface roughness and limited reflecting area discussed earlier are real). Several books have shown plots for different frequencies of $|\rho|$ and ϕ_ρ as functions of elevation angle α without making allowance for changes of ϵ_r and σ with frequency, but account has been taken of the variations of ϵ_r and σ with frequency shown in Fig. 4.6 in plotting Fig. 4.7. Evidently ρ and ϕ_ρ as functions of α do not change significantly with frequency in the range $0\cdot1$ to 3 GHz for moderately dry ground, but do change with frequency for reflection from the sea. This is due to the effect of the permittivity, which is constant in this frequency range, being much greater than that of conductivity for dry ground. Whereas, for sea water, the effect of conductivity is much greater and is very frequency dependent. At very low angles of elevation $|\rho|$ tends to unity for both vertical and horizontal polarisation, and ϕ_ρ tends to 180°, i.e. there is a phase reversal of the reflected wave. However, even for an

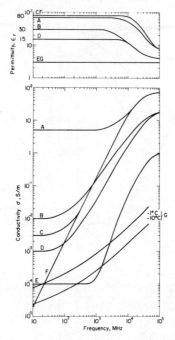

Fig. 4.6. *Permittivity ϵ_r and conductivity σ as a function of frequency*
A Sea water (average salinity), 20°C   E Very dry ground
B Wet ground   F Pure water, 20°C
C Fresh water, 20°C   G Ice (Fresh water)
D Medium dry ground   [Courtesy CCIR[303]]

elevation angle of 1°, $|\rho|$ falls well below unity for vertically-polarised waves reflected from the sea surface; and for elevation angles between 2 and 10°, $|\rho|$ has a minimum value as ϕ_ρ undergoes a maximum rate of change. The corresponding angles of incidence $(\pi/2 - \alpha)$ are referred to as the pseudo-*Brewster angles*. True Brewster angles occur if the surface conductivity is zero, as is almost the case for medium-dry ground, the reflection coefficient being zero at this angle for vertically-polarised waves as the phase lag changes from 180° to zero. By contrast, the reflection coefficient for horizontally-polarised waves scarcely differs from unity for elevation angles less than about 10°, although the difference is more marked over land. Furthermore, the phase reversal condition ($\phi_\rho = 180°$) remains almost constant for all elevation angles. Fig. 4.7 shows that $|\rho_V| = |\rho_H|$ at vertical incidence, although the phase lag shows a 180° difference. Clearly there is no physical distinction between the waves at vertical incidence and the difference in phase is

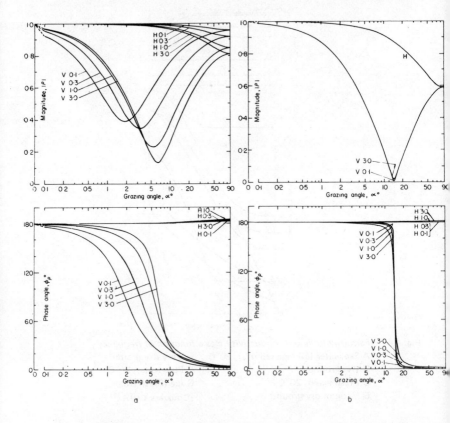

Fig. 4.7. *Magnitude ρ and phase ϕ_ρ of the reflection coefficient of a plane surface as functions of grazing angle α for vertical V and horizontal H polarisations at 0·1, 0·3, 1 and 3 GHz having the following electrical characteristics*
a Sea water
b Medium dry ground

not of practical significance, as may be seen from Fig. 4.8. This shows the phase change to be 180° for horizontally-polarised waves since the electric vector-direction is changed; whereas the phase change is taken as zero for vertical polarisation since the E vector direction is unchanged for $0 < \psi <$ Brewster angle. However, the direction of E

is completely reversed for normal incidence, and no distinction can be made between vertical and horizontal polarisations.

The effect of only a *limited area* contributing to specular reflection is illustrated by Fig. 4.9. It depicts a radio path, TR, above a base plane

Fig. 4.8. *Phase reversal on reflection at conducting surface*

    a Horizontal polarisation: direction of E vector changes from out of reflecting plane ⊙ to into reflecting plane ⊗. Direction of H vector remains constant. Nominal phase change $\phi = 180°$

    b Vertical polarisation: direction of E vector remains constant. Direction of H vector remains constant. Nominal phase change $\phi_\rho = 0°$

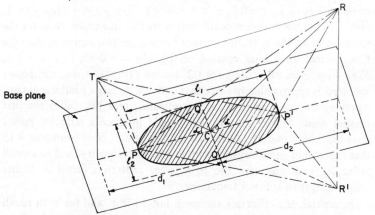

Fig. 4.9. *Plane reflecting surface of limited area equal to nth Fresnel zone*

having a reflecting area bounded by the ellipse PQP'Q', such that the paths TPR, TQR, TP'R and TQ'R each exceed by $n\lambda/2$ the path TCR, where T and R are the path terminals and C is the reflection point. Considering the image point, R' of R, the area bounded by PQP'Q' may be regarded as an aperture corresponding to the nth Fresnel zone in an otherwise opaque screen. Then l_2 is equal to $2r_n$, as given by eqn. 4.1, and l_1 is equal to $l_2/\sin \alpha$. By way of example, if $\alpha \simeq 1°$ then $l_1 \simeq 60 l_2$, and so the distance PP' along the path for which reflection can occur will normally limit the number of Fresnel zones contributing to the reflecting area, rather than the distance QQ′ across the path. For reflection from a surface limited to $l_1 \simeq l_2 \ll (d_1 + d_2)$ the problem is thus effectively one of diffraction at a long slot of width $l_1 \sin \alpha$. In general, the reflecting zone will not extend for $l_1/2$ equally each side of C, but will be asymmetrical. The transmission factor for the aperture, or reflection factor for the surface, is then given by

$$f(\nu) = f(\nu_P) + f(\nu_{P'}) - 1 \qquad (4.12)$$

where $F(\nu_{P, P'}) = -20 \log_{10} f(\nu_{P, P'})$ is the value of the Fresnel-Kirchhoff function given in Fig. 4.2 for

$$\nu_{P, P'} = -\frac{d_{P, P'}}{\sin \alpha} \left(\frac{d}{2 d_1 d_2 \lambda}\right)^{1/2} \qquad (4.13)$$

and where d_P is the distance PC and $d_{P'}$ the distance CP′. By way of examples, if d_P and $d_{P'}$ are zero, i.e. for zero slot width $\nu_{P, P'} = 0$, $F(\nu_{P, P_1}) = +6$ dB, $f(\nu_{P, P'}) = 0.5$ and $f(\nu) = 0$, i.e. there is no transmission. For $\nu_{P, P'} = -0.7$, $(\sqrt{n} \simeq -0.56)$, $F(\nu_{P, P'}) = 0$, $f(\nu_{P, P'}) = 1$, $f(\nu) = 1$, the transmission coefficient is that of free space. Next for the special case of the reflecting surface being a slit just narrower than the first Fresnel zone (and centred on C), $\sqrt{n} \simeq -0.85$, $\nu_{P, P'} = -1.2$, $F(\nu_{P, P'}) \simeq -1.6$ dB, $f(\nu_{P, P'}) \simeq 1.2$ and so $f(\nu) \simeq \sqrt{2}$, i.e. the power reflected is approximately twice that due to a surface of infinite extent. Applying this to a specific case, a reflecting area just filling the first Fresnel zone, PQP′Q′ of Fig. 4.9, would produce a reflected power level four times that of a surface of infinite extent. Note from eqn. 4.13 that ν becomes more negative with increasing frequency, and so a small reflecting area may give rise to a larger reflection factor at higher frequencies than at lower frequencies.

In general, the reflecting surface is not a plane, and for both rough or smooth surfaces the surface curvature may be allowed for by a *divergence factor D*, such that the reflection coefficient R is the product of that for a plane surface and the factor D. For reflection from a smooth earth's surface (assuming a positive earth-radius factor), D is given by the expression

$$D = \left[1 + \frac{2 d_1 d_2}{a_e d \tan \alpha}\right]^{1/2} = \left[1 + \frac{2 d_1 d_2}{a_e (h_1' + h_2')}\right]^{1/2} \qquad (4.14)$$

where α, a_e, d_1, d_2, h_1' and h_2' are as shown in Fig. 4.5. If the effective earth radius a_e is negative, D is greater than 1 and the concave surface causes convergence (or focussing) of the reflected energy, but eqn. 4.14 may still be used. It is noteworthy that D tends to unity for large values of a_e, h'_1 and/or h'_2, but tends to zero for small a_e or $(h'_1 + h'_2)$, i.e. for grazing incidence. Eqn. 4.14 is equally applicable for reflection from any other curvature, e.g. that of a hilltop, that may be approximated by a sphere.

For practical purposes it is often sufficient to assume the ground to approximate to a *large flat surface*, so that R may be taken as -1 for small grazing angles, assuming for present purposes that $|\rho| = -1$, $f(\nu) = 1$, $D = 1$, $f(\sigma_i) = 1$, and $\phi_\rho = 180°$. Then $h'_1 = h_1$ and $h'_2 = h_2$ (see Fig. 4.5). These approximations are particularly useful for short paths over sea at UHF and VHF, and over land at the lower frequency end of the VHF band, where ground irregularities become relatively insignificant. Eqns. 4.8 and 4.7 then give

$$|E| = 2|E_d||\sin(2\pi h_1 h_2/\lambda d)| \tag{4.15}$$

A particular case of importance occurs when $h_1 h_2$ is much less than $\lambda d/2\pi$. Then for an isotropic radiator of power P_t a distance d from an isotropic receiving antenna, the power received P_r is

$$P_r = P_t(h_1 h_2/d^2)^2 \tag{4.16}$$

from eqns. 1.9, 1.10 and 4.15. Eqn. 4.16 is used in Section 7.3.3 when considering mobile systems. It is valid within the regions where the field decreases monatonically with distance, or with terminal height reduction, until the radio horizon is reached.

For VHF antennae close to ground, the *'effective antenna heights'* h_t and h_r for the transmitter and receiver, respectively, must be used in eqn. 4.16 in place of h_1 and h_2 to allow for the effects of the relative permittivity ϵ_r, and the conductivity σ (S/m) of the ground.[129, 130] These effective heights are related to the physical terminal heights above ground level by

$$h_t = (h_1^2 + h_0^2)^{1/2} \tag{4.17}$$

and

$$h_r = (h_2^2 + h_0^2)^{1/2} \tag{4.18}$$

where for vertical polarisation

$$h_0 = (\lambda/2\pi)[(\epsilon_r + 1)^2 + (60\lambda\sigma)^2]^{1/4} \tag{4.19}$$

and for horizontal polarisation

$$h_0 = (\lambda/2\pi)[(\epsilon_r - 1)^2 + (60\lambda\sigma)^2]^{-1/4} \tag{4.20}$$

Values of the ground constants ϵ_r and σ are given in Fig. 4.6. By way of example, if $\lambda = 3$ m ($f = 100$ MHz), $\epsilon_r = 15$ and $\sigma = 10^{-3}$ S/m (i.e. for medium dry ground conditions of Fig. 4.6), then h_0 in eqns. 4.19 and 4.20 will be 1·9 m for vertical polarisation and 0·13 m for horizontal polarisation, so the effective height of an antenna 3 m above ground will be increased to 3·6 m for vertical polarisation but unchanged for horizontal polarisation. Over sea water ($\epsilon_r = 70$, $\sigma = 5$ S/m) $h_0 = 14$ m for vertical polarisation and 0·016 m for horizontal polarisation. At frequencies above VHF the distinction between true height and effective height is immaterial for practical purposes.

Expressing eqn. 4.15 in terms of the clearance height h_c normalised by the first Fresnel zone radius r_1 at the point, as discussed in Section 4.2.1, leads to the relationship

$$|E| = 2 |E_d| |\sin \{\pi(h_c/r_1)^2\}| = 2 |E_d| |\sin (\pi n/2)| \quad (4.21)$$

Furthermore, since the diffraction loss, or attenuation with respect to free space, is $A = 20 \log_{10}(|E_d|/|E|)$dB, eqn. 4.21 may be written in the alternative forms

$$A = -6 - 10 \log_{10} \sin^2 (\pi \Delta_r/\lambda) \quad (4.22)$$

or

$$A = -6 - 10 \log_{10} \sin^2 (2\pi h_1 h_2/\lambda d) \quad (4.23)$$

or

$$A = -6 - 10 \log_{10} \sin^2 (\pi n/2) \quad (4.24)$$

This last relationship is shown in Fig. 4.10. For the ideally-reflecting flat earth case, the signal rises from zero (i.e. $A = \infty$) at grazing incidence ($h_c/r_1 = \sqrt{n} = 0$) to its free-space level at $\sqrt{n} = 0.58$ (where the path length difference $\Delta r = \lambda/6$).

However, in practical cases the amplitude of the reflection coefficient $|R|$ decreases significantly from unity as the grazing angle increases from zero (especially for vertical polarisation, see Fig. 4.7), so that the gain on the free-space condition (at odd-integral values of n) will become somewhat less than 6 dB as the grazing angle increases, and the attenuation peaks (signal nulls at even-integral values of n) become less deep. The way in which A varies with the height of one terminal (h_1 or h_2) is known as the *height-gain* characteristic for that terminal. If the antenna heights are held constant, it also exhibits peaks and nulls with changes of distance.

For long line-of-sight paths, for which the earth curvature cannot be neglected, the path delay Δr and the attenuation A will vary with time if the *refractive index-height gradient changes with time*. If the refractive index gradient were to become much less negative than usual, or even positive, so that a_e became small, eventually the 'earth bulge' could obscure the path, giving a diffraction path rather than a line-of-sight path. The limiting condition for obscuration may be seen from Fig. 4.5 when $h'_1 = h'_2 = 0$. Then, since h_1 and h_2 are both much less than a_e

$$d = d_1 + d_2 = (2a_e h_1)^{1/2} + (2a_e h_2)^{1/2} \qquad (4.25)$$

Clearly this is the condition $n = 0$ and $\nu = 0$ in Figs. 4.2 and 4.10, where $A = 6\,\mathrm{dB}$ for a diffraction edge and $A = \infty$ for a perfectly-reflecting flat earth. In each case the losses will be greater than free space if $\sqrt{n} < h_c/r_1 \simeq 0.6$.

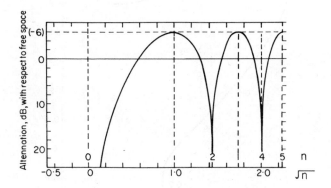

Fig. 4.10. *Attenuation as a function of normalised clearance height, $h_c/r_1 = \sqrt{n}$ when n Fresnel zones are clear of obstruction from a flat, perfectly-reflecting earth*

A further problem of multipath interference encountered on paths over sea is that the phase delay from the reflected path varies with *tidal changes*, and changes in field strength of more than 40 dB have been reported at UHF due to this cause. Typically, a sea-reflected ray at 500 MHz with $1°$ grazing angle and horizontal polarisation will be within 0·1 dB of the direct component. For vertical polarisation the figure would be 3·5 dB below the direct component. Some improvement may be achieved by ensuring that the geometrical point of reflection on the water is screened from one terminal or the other. Indeed, this may be essential to achieve wide bandwidth transmissions (i.e. high data rates). Also, some reduction in the fading may be achieved by height or frequency diversity (see Section 4.6).

In order to select that the *point of ground reflection* on a line-of-sight path has a minimum reflection coefficient, i.e. by ensuring that it is over land rather than water, or over a built-up area rather than over a plain, a means is required for calculating the distance d_1 of the reflecting point from one terminal for a range of values of effective earth radius a_e. For a smooth earth, this can be done in terms of the path characteristics h_1, h_2 and d of Fig. 4.5, but it involves the solution of a cubic equation of the form

$$d_1^3 - \tfrac{3}{2} d d_1^2 + [\tfrac{1}{2} d^2 - a_e(h_2 + h_1)] d_1 + a_e h_1 d = 0 \quad (4.26)$$

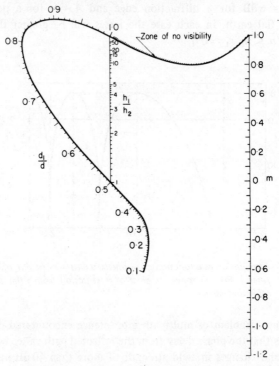

Fig. 4.11. *Nomogram to evaluate the reflection point for a line-of-sight path over a smooth earth*

Gives distance d_1 of reflection point from terminal for terminal heights h_1 and h_2 above datum, path length d and effective earth radius a_e (all in the same units)

$h_1 > h_2$

$m = d^2 / [4 a_e(h_1 + h_2)]$

($m < 0$ if $dN/dh < -157$ N/km)

[Based on Norton *et al.*[132] and Boithias[133]]

Solutions have been proposed in terms of nomograms,[132, 133] of which one is given in Fig. 4.11. If a straight line of alignment cuts the shaded part of the curve marked 'zone of no visibility', the value obtained is not line-of-sight. In the special case of $m = d^2 / [4 a_e(h_1 + h_2)] < -1$, the three roots of eqn. 4.26 are real, i.e. specular reflection may occur at three points along the path. For $m > -1$, two of these roots are imaginary. Fig. 4.11 may be used to examine d_1 as a function of the effective-earth-radius factor k (or a_e).

For undulating terrain, the possible reflecting points may be evaluated in a piecewise manner from equating the surface by a sequence of tilted reflecting planes, though superimposing the Fresnel ellipsoids on the plotted path profile will give a more satisfactory indication of the likely ground reflected component.

4.3 Other multipath propagation losses

The last Section was concerned with multipath losses due to ground reflection and this is often a major problem. However, at frequencies above VHF some account must also be taken of possible reflections from terrain features such as hills, buildings, trees or other obstacles, the smaller obstacles becoming more important at shorter wavelengths. Furthermore, layered refractive-index changes may have a very significant effect.

Just as reflections from ground may be reduced (or even eliminated) by suitable siting of the path terminals, so it may be possible to site the terminals so that local obstacles screen any large *reflections from hills*. Because the direct and reflected paths will be at a similar height, refractive-index changes on the two paths are likely to be comparable, and the interference fading associated with ground reflections is less likely. If a hill is very far off the direct line, it may be possible to use the antenna gain pattern to discriminate against the unwanted reflections.

Reflection from buildings close to one terminal may be a consideration in setting up a line-of-sight microwave link across a city, and existing links may influence the design and orientation of proposed new buildings. The relative importance of specular and non-specular reflections, according to the surface material used, have been considered in a detailed experimental study carried out at 9·4 GHz.[134] When specular conditions were satisfied, the scattered field at distances of a few hundred metres from a building was only a few decibels weaker than the incident field. The scattering with vertical polarisation was generally a few decibels stronger than with horizontal polarisation. The dependence of scattering on azimuth, elevation and range can often be reasonably well explained by the simple theory for smooth flat elements having sizes corresponding roughly with those of actual building features. Corner reflectors formed by adjacent orthogonal surfaces of buildings can cause strong back-reflections and their effects can be reasonably well predicted. Also regularly spaced features such as window frames, can act as diffraction gratings and so lead to preferred directions of scatter. Multiple scattering between buildings may also be significant, particularly since two successive reflections off opposite-facing parallel-oriented buildings will produce a beam parallel to another that does not suffer such reflection. Large belts of trees may cause similar reflection to that of isolated buildings. Signal loss due to reflections from relatively small objects very close to the antenna, e.g. single trees, water tanks etc., may usually be minimised by a small positional change of the antenna.

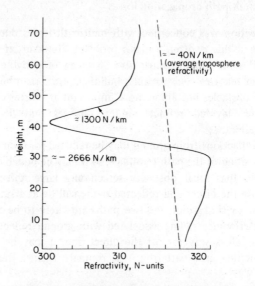

Fig. 4.12.   *Example of a refractive index change with height associated with multipath effects on SHF links*
Measured in UK at 0310 GMT on 2nd July 1975

[Courtesy IEE[135]]

As well as multipath effects due to reflection from ground, buildings or other obstacles near the path, *layered changes of refractive index with height* can disrupt line-of-sight communications in several ways. Consider for example the refractive index-height profile shown in Fig. 4.12, which was recorded in the UK while conducting measurements at 11, 19 and 36 GHz on a 7·5 km path.[135] Such events are rare, but are not negligibly so. The effect on radio transmissions depends very much on the height of the path terminals. If for the example given in the Figure, a transmitter happened to be near 40 m height, then the ray pattern would tend to diverge rapidly with distance, and very considerable transmission loss could occur. In the study mentioned above, it was thought that this mechanism produced a loss of at least 12 dB. If the layer happened to be between the transmitter and receiver heights, partial reflection (at lower frequencies) or refraction (at higher frequencies) could give rise to substantial losses. If the layer were clear above (or below) the transmitter and receiver heights, multipath delays could be introduced by reflection or refraction. In the study mentioned above, a delay of less than 1 ns and another of 6 ns were observed. The latter was associated with a fade of 34 dB. The shorter delay was probably due to refraction, and the longer delay due to total internal reflection. A stratified layer such as is shown in Fig. 4.12 may produce a sequence of multipath effects if its height and/or profile shape change with time. The sharp change with height may exist over a large proportion of the first Fresnel zone of a typical path, i.e. most of its length. On overland paths, these sharp layer discontinuities are particularly associated with mist or fog over moist river valleys or moors,[136] and normally occur during calm cool nights and early morning of hot summer weather (see Section 2.5). A single decrease with height is the more common form. For oversea paths, the reflections may occur at the top of evaporation or advection ducts, since the boundary between the dry air (above) and moist air (below) may be quite abrupt. Models have been developed to show how the use of terminals with sufficient difference in their heights can prevent significant multipath fading, be it on overland or oversea paths.[137, 138]

4.4 Losses due to precipitation, fog and atmospheric gases

In Section 3.4 there was general consideration of the attenuation due to atmospheric gases, fog and precipitation, and how attenuation due to rain is dependent on the rainfall rate, the drop sizes, the radio frequency and the polarisation. These meteorological phenomena are of no consequence at VHF and UHF, but may severely attenuate line-of-sight radio signals at frequencies above UHF.

Measurements of *fading due to rain* on line-of-sight paths have shown that the year-to-year variability is large, especially for small time percentages. Among the many possible examples, on 11 GHz links extending over 10 and 32 km the attenuation that occurred for 0·001% of time in one year occured for more than 0·02% of time in another year.[142] In another study, the time percentage for which a rainfall rate of 40 mm/h occurred in one year was a factor of ten greater than the corresponding time percentage for another year.[59]

Efforts have been made to establish reliable prediction methods by identification of some relationship between the expected attenuation statistics for a path, the path length and statistics of rainfall rate at a point. This is complicated because the distribution of sizes and separations of raincells are to some extent dependent on climate and the rainfall rates even more so. Furthermore, nearly all existing rainfall-rate statistics have been collected either with slow response-time instruments which do not show the intense rainfall rates that may occur for only a few minutes at a time (see Section 3.3), or collected over too few years to include the rarest high rainfall rates.

However, as discussed in Section 3.2, some progress has been made in modelling the spatial distribution of rain cells. As well as providing simplified attenuation statistics, this method facilitates examination of fade durations and the joint probability distribution for links operating in tandem or in diversity mode.[139] Close similarity between recorded and predicted attenuation statistics has been reported for the climatic conditions to which the above methods have been applied. Other studies have involved placing a number of high-performance rain gauges along line-of-sight paths, so that the instantaneous measured attenuation on the paths could be compared with that computed from the rainfall data.[140-142]

From modelling of raincells, some success has been achieved in producing *reduction factors* which may be used to scale the statistics of rainfall rate at a point to produce 'equivalent rainfall rate' statistics which may then be applied to the whole path length. These equivalent rainfall rates may be used together with the path length, the frequency and the specific attenuation versus rainfall-rate relationships of Fig. 3.12 to estimate the attenuation statistics for the path, and approximate agreement with measurements of attenuation on radio paths has been achieved in certain climates.[85, 143-145] However, the factors are particularly inapplicable in climates where intense rain occurs over large areas at a time (e.g. monsoon rain).

Attenuation due to atmospheric gases may be significant at frequencies above about 15 GHz. The attenuation due to oxygen (which is approximately constant with time) and that due to water vapour (which is variable) are shown separately in Fig. 3.14. The attenuation due to *cloud or fog* is shown in Fig. 3.13 in conjunction with Fig. 3.7.

A summary of the specific attenuation due to rain, fog and atmospheric gases was given in Fig. 3.1. However, their relative importance on line-of-sight paths is not immediately apparent, because rain is likely to fill only a small part of a path, and fog too may not fill it completely. By way of example, for a 60 km path at 10 GHz:

(*a*) rain of 25 mm/h over 5 km would produce 0·45 × 5 = 2·3 dB attenuation

(*b*) fog of 1 g/m^3 over 30 km would produce 0·02 × 30 = 0·6 dB attenuation and

(*c*) atmospheric gases over 60 km would produce 0·011 × 60 = 0·7 dB attenuation.

By contrast, for a 20 km path at 35 GHz (a 'window' in the atmospheric absorption curves):

(*a*) rain of 25 mm/h over 5 km would produce 6 × 5 = 30 dB attenuation

(*b*) fog of 1 g/m^3 over 20 km would produce 0·9 × 20 = 18 dB attenuation and

(*c*) atmospheric gases over 20 km would produce 0·07 × 20 = 1·4 dB attenuation.

It may be assumed that such a path would not suffer attenuation from fog and rain at the same time. In each case a water vapour concentration of 7·5 g/m^3 has been assumed. However, in a climate where the air is dry and rainfall is rare, it may be practicable to operate short line-of-sight communication systems at frequencies up to 150 GHz so long as the absorption peaks near 60 and 119 GHz are avoided; indeed, the absorption will not exceed about 10 dB/km in a horizontal path until the frequency exceeds 300 GHz, apart from near the additional absorption line at 183 GHz.

4.5 Refractive fading

For frequencies at which attenuation to rain is not significant, i.e. below about 5 GHz in the present context, the analysis of a large number of radio paths in Europe has shown that the fading depth below the free-space value exceeded for a small percentage of time is primarily a *function of the path length*, as is indicated in Fig. 4.13. These curves give the statistical distribution of the fading depth (relative to free space) for 4 GHz during the worst month of a year for average terrain. As well as the distance dependence there is some secondary dependence on the frequency and type of terrain traversed, and similar statistics may be derived for other terrain and frequencies. At 2 GHz the fading depth exceeded for 1% of time is less than that of Fig. 4.13 by only 0·5 dB for 50 km range and 5 dB for 250 km range; whereas at 6 GHz it is greater than that of Fig. 4.13 by 1 dB for 50 km range and 6 dB for 250 km range. For 0·01% of time, there may be more than 6 dB additional fading for paths of less than 100 km over fairly smooth terrain, and more than 12 dB additional fading over water, moist river valleys or moors.[314]

For designing radio-relay systems it is sometimes necessary to predict the probability of deep fades for very small percentages of the time (e.g. a total of about 0·0002% of a year, or 1 min per year).

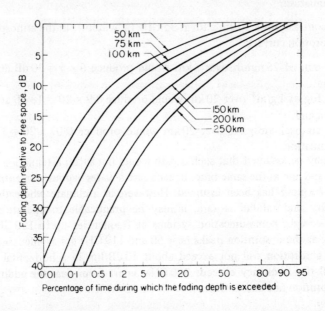

Fig. 4.13.   *Fading depth exceeded for a given percentage of the worst month of the year*
For the path lengths indicated, and at a frequency of 4 GHz, over average rolling terrain in Europe

[Courtesy CCIR[314]]

Measured data are seldom available for such small time percentages, but *Rayleigh-distributed fading* may be assumed,[146] since these very deep fades are due to very many contributions arriving over slightly different paths with different phases. For North West Europe the distribution has the form

$$P(W) = 1\cdot4 \times 10^{-8} \frac{W}{W_0} f d^{3\cdot5} \tag{4.27}$$

where $P(W)$ is the probability that the received power is less than, or equal to, W during the worst month, W_0 is the power received in non-fading conditions, f is the frequency (GHz), and d is the path length (km). The formula applies only for clear line-of-sight paths with negligible earth reflection, and for fading in excess of 15 dB. The multiplying factor in the equation has been found to vary over two orders of magnitude according to climate and terrain roughness.[314]

At frequencies greater than 10 GHz, *scintillation* may be particularly important. This scintillation is caused by fluctuations in refractive index which produce a slight focussing and defocussing of a line-of-sight radio beam. The received signal is then the sum of a large number of components that arrive from different directions with continually changing amplitude and path length, so that the resultant shows a rapid and random variation of amplitude and phase. In the optical case, scintillation is the cause of stars appearing to twinkle and of distant objects appearing to shimmer through a heat haze. In general terms, the standard deviation of the logarithm of the received power σ_s is given by[147]

$$\sigma_s = 19 \cdot 0 \, \lambda^{-1/2} \left[\int C_n^2(d) \, d^{5/6} \, dd \right]^{1/2} \qquad (4.28)$$

where $C_n^2(d)$ is the refractive index structure function (see Section 2.7) at distance d along the path. The effects are reduced if large antennae are used, since the range of angles (i.e. different paths) is then reduced, and there is some spatial averaging across the aperture of the incident phase front.

With paraboloidal antennae up to a few metres in diameter, it is estimated that the peak-to-peak fluctuations occasionally experienced on line-of-sight links with terminal heights of 10 or 20 m will be about ± 5 dB on a 10 km long path at 100 GHz, and ± 3 and ± 8 dB for path lengths of 10 and 50 km, respectively, at 35 GHz. The phase differences due to scintillations which can occur across the wavefront of a large antenna on a line-of-sight path have been estimated from theoretical work by assuming models of refractive index structure. For a 6 m spacing, the path difference for small time percentages may reach 0·25 mm at 2 km, 0·6 cm at 10 km and 1·3 cm at 50 km, corresponding to 10°, 25° and 55° phase difference at 35 GHz. This is borne out by the limited experimental data available. Measurements of the apparent angle of arrival on a 30 km long line-of-sight path have shown rapid changes of ± 45 arcminutes about the line-of-sight,[122] although variation in the horizontal plane is usually less than 6 arcminutes. If for the various contributions to a given signal level there is a range of delay times Δt, then there will be only a limited bandwidth B Hz that can be transmitted without distortion, such that approximately

$$B = 1/\Delta t \qquad (4.29)$$

For multipath fading the range of delay times may be large so that the available bandwidth is small, whereas the converse is true for scintillation. The acceptable bandwidth depends on the acceptable distortion for the service type (e.g. multiplex telephony, television pictures etc.). To transmit rapidly varying (wideband) signals, it may be possible to reduce the range of path delays by the use of narrow beam antennae. For line-of-sight radio links, this bandwidth limitation may be most severe when multipath propagation occurs, and in temperate climates this appears mainly during the night and early morning of summer days.

In addition to knowing the statistics of fading amplitude and time delays, it is important also to know the *fade durations*. If the total time for which deep fades occur were made up of a large number of very short fades, they would cause a rise in error rate of high-speed digital data communication, but not cause an interruption to telephone links. Conversely, if the total time were made up of a few long-duration fades, their interruption to telephone communication might be severe. Measurements suggest that the distribution of the duration of fades approximates a log-normal law. The median value of the fade duration decreases when the fading depth increases and with increasing path length, but increases with greater path clearance from obstacles. The value of the standard deviation decreases when the fading depth increases, but increases with frequency. It depends only slightly on the geometrical characteristics of the path and the climatic conditions encountered.[314] Long-term measurements on microwave line-of-sight paths of 40 to 70 km in the USA have shown that the median duration t_{50} s of multipath fades can be related to the path length d km, the frequency f GHz and the fade depth F dB greater than 20 dB by[148, 149, 137]

$$t_{50} = 56 \cdot 6 \times 10^{-F/20} \left(\frac{d}{f}\right)^{1/2} \tag{4.30}$$

The *rate of change of fade level* is important for the design of switching systems for diversity reception (see Section 4.6). Changes of 30 dB/s have been recorded on a 45 km overland path at 14 GHz in Denmark by sampling at 1 s intervals,[150] and similar results were obtained in the USA at 4 and 6 GHz.[148] The rate-of-change of fade level due to rainfall on the Danish path did not exceed 10 dB/s, and similar measurements in France have shown fade rates to be much lower for rain than for multipath effects.[151]

4.6 Use of diversity

Signal fading may be reduced by selecting the highest level from two or more signal channels which carry the same information, but are taken from separate receivers, so long as the signal levels fade independently, or near so (i.e. they have a low cross-correlation). In most such diversity systems either two receiving channels operate at different radio frequencies (frequency diversity) or they are fed from two antennae spaced a distance apart (space diversity). Such a two-channel system is described as 'double (channel) diversity'. Sometimes two such diversity systems are used together in quadruple-channel-diversity, and such a case is illustrated in Fig. 4.14.

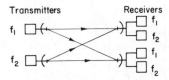

Fig. 4.14. *Quadruple diversity system*
Space/frequency operation

Two parameters used to assess the performance of a diversity system are diversity gain and diversity advantage. The *diversity gain* of a system is the ratio of the power output at a given time percentage to the corresponding power that would be obtained from a single channel. The *diversity advantage* is the ratio of the time percentage for which a given fade level is exceeded for a single channel to that for a diversity system. These two parameters may be derived from cumulative distribution fading statistics of the type shown in Fig. 4.15 for a two-channel system. In the example given, the diversity advantage for a single-channel fade level of 6 dB (mean of two channels) is $p_1/p_2 = 0.085/0.0082 = 10.4$, whereas the diversity gain is $A_1 - A_2 = 6 - 2.2 = 3.8$ dB. In general, both the diversity advantage and diversity gain increase as the level of attenuation being considered increases, or the percentage of time decreases. The use of diversity advantage is attractive when comparing time percentages for which an acceptable system performance might be attained, but diversity gain is less sensitive to year-to-year variations in statistics, and so allows similar conclusions to be reached from data obtained during different observation periods or even different weather conditions.[152]

Before discussing the relative merits of various diversity systems, it is useful to consider the theoretical diversity gain and diversity advantage that an ideal diversity system would achieve. For *Rayleigh fading*, that is where the received signal is made up of many contributions with independent amplitude and phase (i.e. from multipath propagation), signal fades are greater than 18·4 dB below the median for 1% of the time. For two independent Rayleigh signals, if the larger of the two is always selected by a diversity switching system, the switched output fades to only 8·2 dB for 1% of the time, yielding a 10 dB diversity gain.

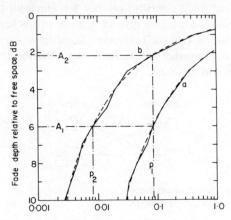

Percentage of time fade depth is exceeded

Fig. 4.15.   *Assessment of diversity advantage p_2/p_1, and diversity gain $A_1 - A_2$ dB from cumulative distribution fading statistics*
   a Distribution for single channel (mean of two channels)
   b Distribution for diversity (best of two channels)

If the best of 4 independent channels were to be selected, the diversity gain would be 16 dB at the 1% level. At the 0·1% level the corresponding values are 15 and 24 dB. The diversity gain of a system also depends on the correlation between the channels. If the correlation coefficient is less than 0·6 on a two-channel system, the diversity gain is 8 dB or better. For a correlation coefficient above 0·6 the performance is steadily degraded. This correlation coefficient of 0·6 may, therefore, be used as a threshold for diversity applications.

In general, the fading will not be strictly Rayleigh-type if there are only a limited number of contributions. Fig. 4.16a shows the effects of a Rayleigh distribution convolved with a log-normal distribution of standard deviation 3, 4, 5 dB etc. on the fading statistics obtained with a single receiver. In parts *b* and *c* the improvement gained by using two and four receivers in switched-type diversity are shown. As a first approximation, for a Rayleigh distribution (shown as $\sigma = 0$), the fade level exceeded for small time percentages increases by 10 dB per decade for a single-channel system, by 5 dB per decade for dual-channel diversity and by 0·25 dB for quadruple-channel diversity.

The diversity advantage at a fading level exceeded for $p_1\%$ of time on a single channel of a system having uncorrelated fading on the signal

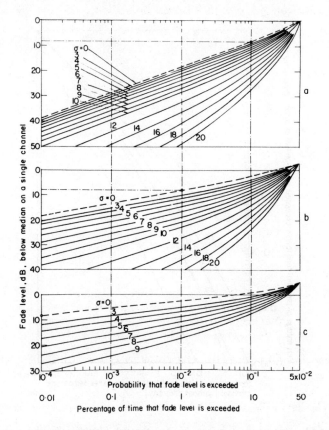

Fig. 4.16. *Fading distribution for convolution of a log-normal distribution of various standard deviations σ with a Rayleigh distribution for (a) a single receiver, and for (b) two, and (c) four independent receiving channels used in switched-type diversity*

—·—·— points mentioned in the text

[Based on Figure from Picquenard[124]]

channels is $p_2/p_1 = p_1 \times 10^{-2}$ for two-channel diversity operation, and $p_1^3 \times 10^{-6}$ for four-channel diversity operation. Fig. 4.16 shows a point of comparison at the 8 dB fade level, where $p_1 = 10\%$, i.e. if the probability of one channel being below the 8 dB level is 10^{-1}, then the probability of two or four independent channels being below the same level is 10^{-2} and 10^{-4}, respectively. However, in practice the signal channels of a diversity system are never completely uncorrelated, and so these diversity advantage figures are not achieved.

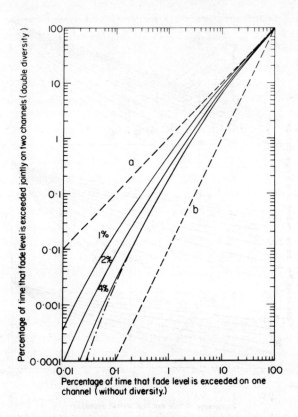

Fig. 4.17. *Improvement gained by two-channel (double) diversity operation on line-of-sight paths during the worst month*

    a Limiting curve without diversity

    b Limiting curve with ideal switched diversity

  ————— Frequency diversity with $\Delta f/f = 1\%$, 2% and 4%

 —·—·—·Space diversity with spacing greater than 150 wavelengths

[Based on Boithias and Battesti[123]]

An indication of the improvement that may be obtained with *space or frequency diversity* on line-of-sight paths is given in Fig. 4.17.[123] The percentage of time for which the highest of the two independent channels of a switched (double-diversity) system might be expected to fade below an indicated level is shown plotted against the percentage of time for which any one channel might be expected to fade below the same level. Clearly, limits are set by the no-diversity curve (curve *a*) for which the percentage of time on both scales are the same, and the theoretically-idealised switched-diversity condition (curve *b*) for which the probability of two independent variations being below a certain level is the square of the probability of either single variable being below that level. The remaining curves on Fig. 4.17 show the measured improvement that may be achieved between about 2 and 10 GHz by space diversity or by different frequency spacings on a line-of-sight path, i.e. different degrees of correlation. The curves apply for the least favourable month on links longer than 75 km seriously affected by multipath fading. The improvement would probably be slightly better for lower frequencies or shorter paths.

In the case of space diversity, vertical separation of the antennae is more effective than horizontal spacing in protecting line-of-sight links against the multipath effects of ground reflection or reflection from an atmospheric layer, and for this reason the technique is sometimes referred to as 'height diversity'. A single antenna may be used at one end of the link, and two appropriately spaced antennae at the other end. The correlation coefficient ρ_s for two such antennae spaced vertically by Δh (m) at one end of a link of d (km) length has been found to be

$$\rho_s = \exp\left[-0.0021\,\Delta h f \sqrt{0.4d}\right] \tag{4.31}$$

where f(GHz) is the frequency.[153] It is assumed in this equation that the ground-reflected wave is negligible. Vertical spacings of 150 wavelengths are often considered appropriate to reduce fading due to multipath propagation, and eqn. 4.31 shows that this corresponds to a correlation coefficient of 0.65 for a 50 km path length.

If ground reflections cannot be avoided, and the ground-reflected component at the receiver is likely to be high (as on over-water paths), this will lead to the most severe form of multipath fading, i.e. two components of nearly equal amplitude. However, a diversity system may be designed for the path, to ensure that when one receiving antenna has a minimum signal strength, there is a high probability that the other has a usable signal. This may be achieved by taking note of the periodicity of the height-gain pattern that may be expected at a terminal site of the link (see Section 4.2.3) and setting the vertical spacing to half the separation of peaks. The signal level on each antenna may then vary due to changes in refraction (and effective-earth-radius factor) on the path, but the combined output from the two receiving antennae will not experience the most severe fade levels.

So far this Section has been considering diversity results for switched-diversity systems, i.e. picking out the instantaneous peak of two or more channels of data. However, if a phase-combining diversity technique is used rather than a switched-diversity technique, only one receiver is required instead of two, and the breaks in the received signal due to switching are avoided.[154] In this system the signal from the two antennae are aligned in phase and added together in a waveguide directional coupler.

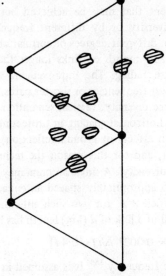

Fig. 4.18. *Route diversity on a microwave communications network system, with rain attenuation on three links, but full service maintained*
    ⬡ Raincells
    ● Link terminals (typically separated by 10 km in 20 GHz system, or 30 km in 11 GHz system)

At frequencies where attenuation due to rainfall is significant on line-of-sight links, e.g. above about 5 GHz, protection against fading may be obtained using *route diversity*. The term 'route diversity' is used rather than space diversity because the link terminal spacing is large enough to ensure that rain on one arm of a diversity system is unlikely also to affect an alternative arm. This system can be provided in a microwave communication network, such as that illustrated diagrammatically in Fig. 4.18. In this instance, by suitable route switching a full service is maintained, although three of the links are lost temporarily due to rain attenuation. The design parameters of such a system depend largely on the characteristics of raincell sizes and separations mentioned in Section 3.2, but results from a study in the UK indicate that a separation of 4 km of parallel paths gives a considerable improvement in a switched-path system.[155, 156] Larger spacings do not provide a proportionately greater improvement.

4.7 Use of orthogonal polarisation

Re-use of the frequency spectrum by operating both on horizontal and vertical polarisations has been employed for some time in the broadcasting services at VHF and UHF, but in that application the object has been to obtain extra protection against interference from a distant transmitter on the same frequency. A similar technique may be used to double the frequency capacity of line-of-sight microwave links, using either orthogonal linear or circular polarisations (of opposite rotation). A limit to this technique is that in propagating through the atmosphere, some of the energy transmitted in one polarisation state can be transferred to the orthogonal polarisation state, thus causing interference (or cross-talk) between the two channels. This effect is usually referred to as cross-polarisation. The ratio of the wanted (co-polar) and unwanted (cross-polar) received power is usually referred to as the *cross-polarisation isolation* (XPI). It is the reciprocal of the 'depolarisation factor', usually expressed in decibels, and is positive. The term 'depolarisation' is sometimes used in place of cross-polarisation, but more often it is taken to mean simply a change from the original polarisation state. It is often convenient for test measurements to be made by transmitting on one polarisation only, and receiving on the orthogonal co- and cross-polarisations. The ratio of the co-polar to cross-polar power levels is then referred to as the *cross-polar discrimination* (XPD), but this is equal to the XPI if rain rather than multipath propagation is the cause of signal deterioration.

Cross-polarisation may be caused by multipath propagation or by rain or other hydrometeors on the path. In addition, some cross-polarisation will occur in the *antennae* themselves, which produces a base level of cross-polarisation for a particular system. During normal propagation conditions, the XPD due to the antenna might be typically 35 dB, but experiments carried out at 6 to 13 GHz on paths of 50 km or more have shown that the XPD between horizontal and vertical polarisation could decrease to less than 15 dB for 0·01% of time in adverse weather conditions.[157-159] This decrease may be due to a combination of tropospheric effects and antenna characteristics. In particular, if the antenna has high isolation (or XPD) over only a narrow angle, then any off-axis rays due to *multipath propagation* may produce a significant cross-polar component at the receiver, and this may be severe even when the fading due to multipath is only slight.[160] Furthermore a signal component on the wanted polarisation may be diminished by destructive interference from multipath reflections, while at the same time the unwanted polarisation that arises from the same multipath reflections is enhanced. The field strength of the two polarisations may vary rapidly and be almost uncorrelated, but this fading may be minimised by the use of space diversity. Another cause of cross-polarisation by multipath propagation may be the rotation of the plane of polarisation of rays reflected from atmospheric layers, or from surfaces of terrain features, when the direction of polarisation is other than parallel or normal to the reflecting surface.[125, 157] It is significant that such cross-polarisation may occur without appreciable signal fading.

Cross-polarisation on a path carrying orthogonal polarisation also normally occurs during *intense rainfall*. It is caused by the individual drops falling as oblate spheroids and producing a different attenuation and phase shift for waves polarised parallel or perpendicular to the major axes of the drops (see Section 3.3). In general, any linearly- or elliptically-polarised wave may be described in terms of elements parallel and perpendicular to any axis transverse to the wave, and so the differential attenuation and phase shift as a wave passes through rain will cause depolarisation, and, for two orthogonally-polarised waves, there will be cross-polarisation. Cross-polarisation will not occur for two linearly-polarised waves when all the raindrops have their axes in one of the polarisation planes. In practice this is most closely approximated for vertical and horizontal polarisations, since the raindrop axes tend to be vertical. However, there is often a localised mean canting of the raindrops due to wind shear, and usually a considerable spread of canting angles about the mean due to turbulence, so that considerable cross-polarisation occurs along a path through rain. A detailed study has been made of this effect for rainfall rates up to 150 mm/h and frequencies up to 100 GHz.[104-106] Differential phase shift appears to be the dominant factor in rain-induced cross-polarisation at frequencies below 10 GHz, (and significant cross-polarisation is then possible even for low values of attenuation), whereas differential attenuation becomes increasingly important at higher frequencies.

Various experimental measurements have shown that cross-polar discrimination D (dB) decreases with increasing co-polar attenuation due to rainfall A (dB),[161-167] but the relationship is not yet firmly established. For small elevation angles, and for linear polarisation in a plane $\tau°$ from the vertical, it appears to have the form[314]

$$D = 30 \log f - 20 \log A - 20 \log(\sin 2\tau) \qquad \text{dB} \qquad (4.32)$$

where $8\,\text{GHz} < f < 20\,\text{GHz}$, $15\,\text{dB} < D < 40\,\text{dB}$, and $10 < \tau < 80°$. For vertical or horizontal polarisation, i.e. $\tau = 0°$ or $90°$, the last term is replaced by $+12$. For circular polarisation, D is about 12 dB lower than for vertical or horizontal linear polarisation, or the same as for linear polarisation in a plane inclined $45°$ from vertical, i.e. the last term in eqn. 4.32 disappears. For paths passing within 100 m of ground level, there may be considerable canting of rain drops, and variation in this canting, due to local wind shears. In such circumstances the cross-polar discrimination for vertical and horizontal polarisation will be less favourable than eqn. 4.32 would indicate.

It has been shown that eqn. 4.32 is only weakly dependent on the drop size distribution.[164] For frequencies less than about 8 GHz, attenuation values are very low, but cross-polar discrimination may be expressed as a function of rainfall rate and path length.[106, 168] A line-of-sight path through *snow or ice cloud* in the absence of rain may also experience a marked decrease in cross-polar isolation. This may have a significant effect on terrestrial links in some climates, and it is certainly of conseqeunce on some earth-to-space links where depolarisation above the $0°\text{C}$ isotherm has more influence than rain below. This effect is considered in Section 5.5.

As to the relative importance of multipath effects and rain in reducing D, measurements at 12 GHz on paths shorter than 20 km in Western Europe have shown rain to be more significant,[140] whereas similar measurements on paths longer than 50 km have shown multipath effects to be more significant.[158] However, even though one effect may be dominant, both rain and multipath effects will normally have to be accounted for in any system design.

Chapter 5

Earth-to-space paths

5.1 Introduction

Many of the factors which influence terrestrial line-of-sight link reliability also influence earth-to-space link reliability, especially if the elevation angle from the earth station to the space station is little above the horizon. However, because the earth-to-space link is usually more severely limited in its power budget, propagation problems may be more critical to system reliability. Although some earth-to-space links require only narrow band systems that operate at VHF or UHF, most satellite communication systems operate at SHF, and may soon operate also at EHF, in order to achieve maximum information carrying capacity. This relative priority will be reflected in the content of this Chapter.

At low elevation angles, as when operating between a geosynchronous satellite and an earth station situated at a high latitude, or in maintaining communication with a non-geosynchronous satellite for the maximum time that it is above the horizon, ground reflections, variations in refraction and ducting may all be significant. In addition, radio signals which traverse the ionosphere may be subject to scintillation, absorption, variation in the direction of arrival, propagation delay and changes of polarisation due to the presence of free electrons and the earth's magnetic field. These effects often produce a considerable disturbance of earth-to-space communications. When operating well above the horizon the propagation problems are generally less severe. By way of example, for a geostationary satellite on the same meridian as an earth station at $50°$ latitude, the elevation angle will be $33°$, but this will decrease if the earth station is at a higher latitude (i.e. $22°$ elevation at $60°$ latitude), or if the satellite is not on the meridian of the earth station. The prime propagation problem on earth-satellite links in frequency bands above UHF is attenuation due to precipitation and atmospheric gases.

5.2 Refraction

The extent to which a steady decrease of refractive index with height causes the bending of radio rays was described in Section 2.4. For a radio path extending throughout the atmosphere, refraction causes the elevation angle of a ray at the ground to be greater than if the atmosphere were not present, and in this case the raybending and elevation angle correction are the same. Fig. 5.1 gives the computed relationship

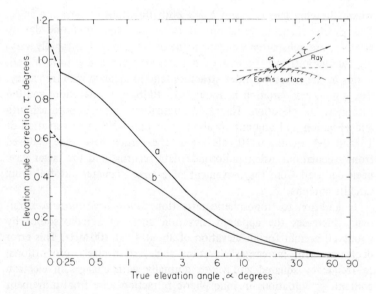

Fig. 5.1. *Error in angle of elevation (ray bending) due to tropospheric refraction*
a Tropical maritime air (July)
b Polar continental air (April)

[Based on data from Shulkin[169]]

between the elevation angle correction and the true elevation angle for a slant path throughout the atmosphere in the extreme conditions of tropical maritime regions (Puerto Rico) in July, and polar continental conditions (Alaska) in April.[169] The curve for median raybending in the UK lies close to the mean of those shown in the Figure. For many applications, raybending may be regarded as negligible for all but small angles of elevation, and most narrow-beam satellite earth stations use an automatic tracking facility which will follow rapid changes of refraction. However, errors less than 0·1° may still be significant for certain radar applications, and errors less than 0·01° are important for some radio-astronomical work.

Slowly changing *variations in refraction* from the long-term median occur seasonally, diurnally and with changes in weather conditions. In certain locations it is possible to establish a high correlation between raybending τ and surface refractive index N_s, leading to the relationship[170]

$$\tau = a + bN_s \qquad (5.1)$$

where the coefficients a and b are both functions of elevation angle. The slowly changing variations in refraction, e.g. from one day to another, are substantially independent of frequency, but rapidly varying fluctuations over periods of minutes or seconds due to local variations in the refractive index structure tend to increase with frequency. The day-to-day variation is about 0·1° RMS at 1° elevation and 0·4 arcmin at 10° elevation. The rapid fluctuations occur in both azimuth and elevation and amount to about 1 arcmin RMS at an elevation of 1° and 0·4 arcmin at 10° elevation. The former have been deduced from calculations using radiosonde data records, and the latter have been deduced from the movement of a large diameter autofollowing satellite antenna.[171]

In addition to tropospheric refraction, *ionospheric refraction* normally increases the apparent elevation angle of a radio source by about 10 arcmin for an elevation of about 4° at 100 MHz. This error decreases with increasing elevation angle, is inversely proportional to frequency squared and varies diurnally with changes in electron content.[172] Variations in ionospheric refraction arise from movement across the line-of-sight of medium- or large-scale ionospheric irregularities. These irregularities, which range from tens to hundreds of kilometres in size, are frequently associated with travelling disturbances and give rise to fluctuations in the direction of arrival of trans-ionospheric signals with periods from about 15 min up to an hour or more. They are most commonly observed in daytime. The magnitude of the fluctuation is typically of the order of 3 arcmins at 100 MHz (sometimes as high as 20 arcminutes[312]) for a 30° inclined path, and is inversely proportional to the square of the frequency. Small-scale ionospheric irregularities, which produce amplitude scintillation, also give rise to angular scintillation. In middle latitudes the magnitude of these is typically 1 arcmin at 136 MHz,[173] but larger values may be expected in the equatorial and high-latitude zones of stronger scintillation.

5.3 Attenuation and fading

5.3.1 Losses due to precipitation, fog and atmospheric gases

The transmission loss on an earth-space path is greater than that for free space due to a number of factors. Some of these (multipath propagation and scintillation) cause relatively rapid fading (e.g. within seconds), whereas the attenuation due to absorption by atmospheric oxygen is permanent, that due to water vapour changes slowly (e.g. within hours), and that due to rain and cloud will also be relatively slowly changing (e.g. within minutes). It is desirable to consider separately the rapid fading and its causes, and the slower fading and its causes, because their consequences for radio-system planning, and the diversity systems used to minimise their effects, are different. Also, their relative probability of occurrence is dependent on climate.

The attenuation A dB due to the *atmospheric gases* oxygen and water vapour on any path is given by eqn. 3.9, but for an inclined path through the whole atmosphere it is usually more convenient to use a model of the form

$$A_g = \gamma_{o_o} r_{e_o} + \gamma_{o_w} r_{e_w} \qquad \text{dB} \qquad (5.2)$$

where it is assumed that the values of the specific attenuation of oxygen and water vapour at the earth's surface γ_{o_o} and γ_{o_w}, respectively, may be applied over effective distances r_{e_o} and r_{e_w}, respectively. At mid-latitudes the decrease with height of specific attenuation due to oxygen and water vapour may be approximated by an exponential function with scale heights (i.e. effective distances for a vertical path) of 4 km and 2 km, respectively, although these scale heights are not very appropriate for frequencies close to absorption lines. Theoretical values of the one-way attenuation for a vertical path through the atmosphere as a function of frequency are indicated in Fig. 5.2 for a median humidity (in temperate climates), curve *a*, and for dry air, curve *b*. The curves are consistent with measured data. Linear interpolation may be used between curve *a* and curve *b* (and linear extrapolation beyond curve *a*) to determine the attenuation for values of water vapour concentration other than $7 \cdot 5$ g/m^3. The curves do not take account of any possible additional attenuation due to the presence of molecular complexes of water, which may be important at frequencies greater than 100 GHz, particularly under conditions of high relative humidity. These effects appear to be very variable.[174],[175] The attenuation for any elevation angle greater than about 5° may be calculated from Fig. 5.2 or eqn. 5.2 by multiplying the

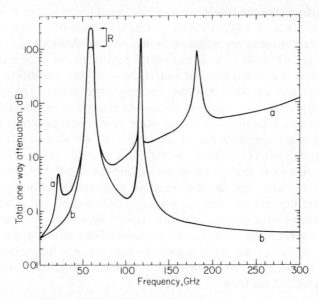

Fig. 5.2. *Total one-way zenith attenuation through the atmosphere as a function of frequency*
a 7·5 g/m³ at ground level
b Dry atmosphere (0 g/m³)
R: Range of values due to fine structure

[Courtesy CCIR[322]]

zenith attenuation path by the cosecant of the elevation angle. The limits to the variability of water vapour concentration at ground level and the variability with geographical location should be taken into account (see Section 2.1).

The occurrence of intense *rain* on slant paths may cause considerable attenuation, but in addition to the general considerations of statistics of rainfall rate measured at a point and the specific attenuation due to rain (as discussed in Sections 3.3 and 3.4), allowance must be made for the spatial non-uniformity of rain. This influences the elevation angle dependence. The most intense rainfall that produces high attenuation is very localised (see Section 3.2), and at high angles of elevation it is unlikely that more than one raincell will be traversed by an earth-satellite path. At low angles, the probability of the path intersecting more than one cell becomes important. An empirical approach to the problem uses an 'effective path length' along which rain of uniform characteristics would cause the same attenuation as that which occurs on the actual path. Because the more intense rainfall tends to occur in smaller cells than does the less intense rain, the effective path length corresponding to a given elevation angle is a function of rainfall rate. Empirical relationships are shown in Fig. 5.3, together with two curves

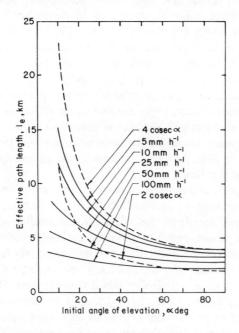

Fig. 5.3. *Effective path length on a slant path as a function of rainfall rate and elevation angle*
Also two arbitrarily chosen cosecant relationships
[Based on CCIR[317]]

for a cosecant law. Such a law would apply for horizontally-stratified rain, but not for vertical columns of rain, and it may be seen that the cosecant law is more appropriate for a rainfall rate of 5 mm/h than for 100 mm/h. Alternative curves are available for climates where high rainfall rates are more widespread (as is the case for monsoon-type rain).[317] This approach to estimating attenuation is still somewhat unsatisfactory and applies only to certain climates. However, for these climates

(*a*) point rainfall rate statistics from Fig. 3.8,
(*b*) curves of specific attenuation versus rainfall rate from Fig. 3.12 and
(*c*) curves of effective distance versus rainfall rate and elevation angle from Fig. 5.3 may be used in the absence of local information as a means of determining the attenuation on a slant path.

Attenuation due to *water cloud or fog* can be calculated from Fig. 3.13 if the liquid water content and physical dimensions are known. Unless clouds have high water content, the total attenuation that they produce is not great, even in frequency bands above UHF. Measurements in the UK showed that at 95 and 150 GHz the attenuation in the vertical direction due to large cumulus clouds was only 1·5 and 2 dB, respectively, and for nimbo-stratus (rain) clouds the attenuation was 2 to 4 and 5 to 7 dB, respectively.[324] For widespread cloud the attenuation may be expected to vary approximately as the cosecant of the elevation angle. Because of the difference in dielectric properties, *ice clouds* give attenuation about two orders of magnitude smaller than water clouds of the same water content for frequencies up to about 35 GHz. At higher frequencies the attenuation due to ice clouds may be significant.

Turning now to *means of measuring attenuation statistics*, it is particularly difficult to measure reliably the probability distribution of the deep and slow fades that occur for such small time percentages as 0·01% (i.e. about 4 min) of a month, a statistic required in system planning, since for this short time the amount of rain (the dominant factor at frequencies above 10 GHz) is very variable from year to year and from place to place within a climatic zone. Measurements of satellite transmission over 10 or 20 years are required from many locations, but so far such measurements have been obtained at only a few locations and for relatively short time periods. These include observations at 4, 13, 15, 18, 20 and 30 GHz.[317]

Because of the very limited opportunities for direct measurement of attenuation on earth-satellite paths at frequencies in excess of 10 GHz and the limitations of conventional radar techniques for this purpose[176] (see Section 3.3), much use has been made of data obtained using sky-noise and sun-tracking *radiometers* at these frequencies. Sky-noise radiometers use the principle outlined in Section 3.6. The measured noise power is an indication of the attenuation that would occur on a radio path through the atmosphere. Sun-tracking radiometers determine the attenuation directly assuming the sun to be a source of known noise power. When used without correction for scattering, the sky-noise radiometry technique tends to underestimate the true attenuation, especially at higher frequencies.[43] For example, at a frequency of 37 GHz with elevation angles of 10° to 30° in a temperate climate, the error was found to be about 10 to 15% (dB) for vertical polarisation and 30 to 35% for horizontal polarisation.

Two apparent advantages of using sun-tracking radiometers rather than measuring sky emission are

(a) that a much wider dynamic range of attenuation can be covered and
(b) that the measurements will be more sensitive to small intense raincells of the type that will cause severe losses on earth-satellite links.

The second point arises because, whereas a sky-noise radiometer using a small antenna will have a wide beam, the diameter of the first Fresnel zone at distance d from the antenna of a sun-tracking radiometer (or satellite-signal receiver) will be approximately $\sqrt{\lambda d}$ (e.g. 19 m at 3 km range for 10 GHz). However, system planners are usually prepared to accept no more than about 5 dB loss due to rain, and this range can be measured by sky-noise radiometers. Also, statistics derived from solar radiometry data may have bias due to the angle of elevation being associated with certain times of day and due to no data being available at night. Intense rainfall associated with thunderstorms which cause severe attenuation is not found to be uniformly distributed over the day, but tends to occur as convective activity develops in the afternoon, or as the air cools at night. Measurements made using sky-noise radiometers alongside receivers of transmissions from a satellite at 15, 20 and 30 GHz, have shown agreement within 1 dB for attenuation less than 10 dB, so that much reliance may be placed on statistical information derived from such radiometers.[176, 177] It must be emphasised that the radiometer time-constant must be kept low if reliable measurements are to be achieved. As with the measurements that have been obtained directly from satellite transmissions (mentioned above), data obtained from various parts of the world using both sky-noise radiometers and sun-tracking radiometers show a very large variation in the attenuation for a given time percentage.[317] This variation is indicated in Table 5.1, where attenuation data are displayed in terms of frequency and time percentage. No account has been taken of path elevation angle or climate, but all the data are from paths with elevation angle greater than 20°. The effect of atmospheric noise temperature on earth-satellite paths is considered in Section 8.2. Clearly this noise causes the signal/noise ratio on such a path to deteriorate more than would be estimated from considering attenuation alone.

System planning sometimes requires data for a '*worst month*' which may be defined variously as the 30-day month or calendar month in a year or period of years having the lowest mean or otherwise specified signal level. Unfortunately, few measurements have produced attenuation statistics which are sufficiently comprehensive to enable valid 'worst month' data to be derived. Moreover, even at a given frequency

Table 5.1 *Attenuation (dB) for certain time percentages deduced using fixed-elevation radiometers at elevation angles greater than 20°*

Frequency band	Number of measured distributions	Percentage of time attenuation was exceeded		
		0·01	0·1	1·0
GHz				
35–37	4	20–37*	6–22	2–8
30	1	25	10	4
20–23	3	9–40*	6–15	2–4
16–19	2	12–13	3–7	2
12–13	4	3–13	2–6	2
9–11	6	4–18	2–6	2

Based on CCIR data[317]

The range of values shown for each frequency band and time percentage is the range observed from the indicated number of measured distributions. No account has been taken of angle of elevation or climatic region for which measurements were obtained, or the period of data collection.
* extrapolated from reported data curves

the results will depend very much on elevation angle, climate and the relative importance of effects due to rain and scintillation or multipath fading in clear air for that climate. However, some such data have been collected using radiometers for about 3 years at 10 sites in Western Europe at 11 GHz at elevation angles between 20 and 40°, and 31 distributions representative of the worst month in a calendar year have been prepared.[178] The data are particularly relevant to proposals for direct broadcasting from satellites to domestic antennae. Year-to-year variations in the distribution of attenuation were found to be comparable with variations with location, and much larger than any systematic variation with elevation angle. No elevation angle dependence could be found in the worst-month attenuation values. However, an elevation dependence may be expected at elevation angles below 20°. A comparison of the results from this study with long-term values for West Europe at 11 GHz indicates that the percentage of time for which a given attenuation is exceeded in an average year is about one-sixth of the percentage of time for which the same attenuation is exceeded in the worst month of the average year.[317] The results also show that the ratio of the percentage of time in the worst month to the percentage of time in the year varied between 10 and 3 for fade levels exceeded for less than 1% of the worst month.

Measurements of *fade duration* have been obtained in the UK at 19 and 37 GHz using sun-tracking radiometers.[179, 182] Heavy rain was found to be the dominant factor. At 19 GHz the maximum fade duration at the 10 dB fade level was 8 min, and at the 5 dB level the maximum was 38 min. At 37 GHz, one fade exceeding 10 dB lasted for more than 27 min, and of those exceeding 5 dB, six lasted more than about 50 min.

5.3.2 Refractive fading

Time-lapse records of signal level of earth-satellite paths sometimes show *scintillation* (i.e. rapid fading of the signal amplitude over a narrow peak-to-peak range of the type referred to in Section 4.5). There is then a range of angles over which energy is received, and the wave front intercepted by the aperture of a large antenna has phase variations (or a range of time delays) which cause a degradation (or coupling loss) in effective antenna gain. This aperture-to-medium coupling loss decreases with elevation angle and increases with frequency but, measurements made in Japan at SHF suggest that the losses will be less than 0·5 dB at 5° elevation angle.[181, 182] As well as rapidly varying amplitude of the resultant of the many ray components, scintillation includes rapid variations of resultant phase, polarisation and apparent angle of arrival (the last point was considered in Section 5.2).

Tropospheric scintillation is unlikely to produce severe fading on earth-satellite links operating at elevation angles above 10° and at frequencies below about 10 GHz. Few measurements have been made at these frequencies of such (rapid) scintillation fading using satellite transmissions, but some are available from North America and North West Europe.[317] These show that scintillation at 20 and 30 GHz at an elevation angle of 20 to 30° in a temperate climate may be of 1 dB (peak-to-peak) for clear sky conditions in summer, 0·2 to 0·3 dB in winter and 2 to 4 dB (very occasionally 6 dB) in some types of cloud. The marked scintillations observed in some cloudy conditions often have a typical fluctuation frequency of 0·5 to 1 Hz, but frequency components up to at least 10 Hz have been observed. Much slower fading, which has been observed with periods of 1 to 3 min is probably caused by relatively large-scale changes of refractive index.[317]

At lower angles of elevation, there is a marked increase in the magnitude of the relatively rapid scintillation fading. Theoretical and experimental studies of tropospheric scintillation made in Japan[181] indicate that the standard deviation of the received power σ(dB) as a function of elevation angle $\alpha°$, between 3 and 20°, follows the relationship

$$\sigma = 6.5 \alpha d^{-1.5} \qquad (5.3)$$

for frequencies between 2 and 10 GHz and antenna diameters between 20 and 40 m. Measurements in the USA[183] show scintillations to have a third of this amplitude, although the spread of the results relative to the mean amounted to factors of more than three. For data collected at elevation angles between 1 and 20° at 4 to 30 GHz the most severe peak-to-peak scintillation has been found to be about $0.5/\sin \alpha$(dB).[317]

At elevation angles below about 20° there is also often a type of fading similar to that observed on terrestrial links, with sudden deep fades (of the order of 20 dB and lasting a few seconds or sometimes much longer) which are indicative of *multipath propagation*. From measurements made in Canada at 6 GHz with an elevation angle of 1°, it was estimated that all but 10% of the fades exceeding 20 dB would last less than about 11 s, and of those exceeding 12 dB, all but 10% would last less than about 35 s. The rate of change of signal level exceeded 1 dB/s for 1% of time and 4 dB/s for 0.1% of the time.[184] Such information on rate of change of signal level is necessary for the design of transponders on satellites intended for multiple access by several earth stations, and for designing diversity systems using switching techniques.

In principle, there may be an additional loss at low elevation angles in a clear atmosphere due to *spreading* of the beam as it passes through the atmosphere.[171, 181] This is caused by the variation of refractive index with height. In practice the loss is less than 0.4 dB for an elevation angle of 3°, and it is zero for vertical incidence. The loss is independent of frequency, but is small compared with the attenuation due to the other causes mentioned above. This effect is to be distinguished from gain degradation, which was mentioned earlier in this Section.

5.3.3 Ionospheric effects
Further limitations on earth-space link reliability may be due to *ionospheric absorption*. Normally this decreases with increasing frequency[185] and may be ignored for frequencies above 70 MHz in equatorial and temperate regions. However, in addition to the generally predictable

variations of the ionosphere associated with cyclic solar activity and seasonal changes, there are less-predictable variations arising from temporary changes in solar activity. These variations have their greatest effect near the auroral zones and over the polar caps, and should be allowed for in planning a service in such regions. At VHF the attenuation is small and decreases rapidly with increasing frequency f, being proportional to $(\sec i)/f^2$, where i is the angle of incidence at 100 km height. For all but the lowest angles of elevation at ground level α, it is convenient to replace sec i in this expression by cosec α. For a path elevation angle of 30° and frequency of 100 MHz, the absorption may reach 4 dB in polar cap regions, 5 dB in auroral polar cap regions, but less than 1 dB at mid-latitudes.[312]

Whereas short-term fading due to tropospheric scintillation is not likely to be important for earth-to-satellite paths at frequencies below 10 GHz, *ionospheric scintillations* of amplitude, phase, polarision and direction-of-arrival may represent practical limitations to space communication systems at frequencies below about 6 GHz, especially in equatorial and auroral latitudes. Normally the electron content of the ionosphere is more or less homogeneous over small horizontal distances, and so does not produce signal fluctuations, but sometimes 'spread F', 'sporadic E' or local perturbations in the electron content caused by solar disturbance may produce scintillations on radio signals transmitted between satellites and the earth. These scintillations are observed sporadically, and for frequencies above 100 MHz their magnitude normally decreases with frequency.[312]

The amplitude of scintillations is moderately severe at high latitudes, but it is a maximum in the geomagnetic equatorial region. In each case it is most severe in the early part of the night. Scintillations in excess of 35 dB have been observed at 136 MHz,[186] and fading of 20 dB has been reported at UHF and SHF.[187] At VHF the fading period may be as long as several minutes, whereas at SHF the fading period varies from 2 to 15s.[312] Scintillation at mid-latitudes is rare, but it generally shows a maximum near midnight, and sometimes, especially in summer, a second maximum around mid-day. A study in Massachusetts and Greenland of peak-to-peak fading levels on trans-ionospheric paths at 136 MHz showed fades in excess of 13 dB occurring for significant time percentages, but the effect was most pronounced at the higher latitude.[188, 312] Models have been made to describe the probability density function of fading due to ionospheric scintillation at VHF and UHF.[189] Time and space diversity schemes may be used to overcome fading during periods of ionospheric scintillation.[312]

5.4 Use of diversity

Some general remarks about diversity schemes were made in Section
4.6 before considering the specific requirements for terrestrial line-of-
sight paths. The present Section is concerned with diversity schemes
on earth-satellite paths, where the loss of signal may be due to scintillation or to the longer period fades which arise from rain cells (or
clouds) passing through the beam. The relative importance of overcoming shorter- or longer-period fading depends on the communication
service in question.

Most application of diversity techniques on earth-satellite paths
has been to reduce longer-period fading using *site diversity*. This technique is useful in overcoming the serious effects of rain on earth-
satellite paths at SHF because the most intense rain occurs in cells
of quite limited extent. Fig. 5.4 indicates as a function of site spacing
the time percentage $p_2\%$ for which attenuation A experienced at
a single site for a given time percentage $p_1\%$ (as indicated in Table 5.1)
is exceeded simultaneously at two spaced sites. It is based on data taken
from cumulative distributions of the form shown in Fig. 4.15 from the
USA, UK and Japan, and no allowance has been made for the elevation
angles or frequencies of the radio paths. Somewhat different data
would be expected from, for example, a monsoon climate, where
intense rain is more widespread. The curves drawn to fit the data
are necessarily only an approximation, but certain general points are
illustrated. For very small spacing, p_2 becomes equal to p_1, since the
same fading pattern will occur on each path, but for spacings of a few
hundred metres there will be some decorrelation of the two fading
patterns so that p_2 will be slightly less than for attenuation due to
cellular rain. When the separation is comparable with a typical cell
size, p_2 will be considerably less than p_1. However, even at large
spacings p_2 will be somewhat greater than $p_1^2/100$ where both paths
are within the same general weather system. This is true because the
two sites cannot be regarded as being statistically independent. Rain
occurring on one path implies an increased probability of rain on the
second path compared to times when there is no rain on the first path.
$p_2/100 = (p_1/100)^2$ would apply if the two were completely independent. For levels of greater attenuation (smaller time percentages
of occurrence) the rate of decorrelation with distance is greater (as
indicated in the Figure) since the effective cell sizes are smaller (see
Section 3.2).

This last statement is equivalent to saying that the *diversity advantage* (as defined in Section 4.6) increases with site separation. For

example, at 20 GHz an attenuation of 10 dB may occur for about $p_1 = 0.1\%$ of time at a single site (according to Table 5.1), whereas the joint probability $p_2\%$ of the attenuation at both of two sites separated by 5 km being below this level is 0.03%, giving a diversity

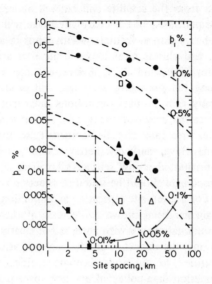

Fig. 5.4.   *Relationship between diversity advantage* p_1/p_2 *and site spacing*
p_1 = time percentage for fading to a given level A (dB) on a single site
p_2 = time percentage for fading to this level A (dB) simultaneously on two sites
Data points:
● Japan   17–35 GHz[190, 191]
○ UK   30 GHz[192]
▲ UK   37 GHz[180]
△ New Jersey, USA   16 GHz[193]
■ UK   12 GHz[194, 195]
□ Texas, USA   30 GHz[196]
—·—·—· points referred to in text

[Based on CCIR[317]]

advantage of 3.3. Similarly, the joint probability for a spacing of 20 km is 0.01%, a diversity advantage of 10. Clearly, economic factors will determine whether the diversity advantage for a 5 km spacing is worth establishing a second site and whether the 20 km spacing is practicable. If diversity is to be considered, the site separation should be greater than the diameter of typical raincells, but less than the typical separation distance of raincells (see Section 3.2).

Generally, the best diversity advantage is gained by sites spaced normal to the earth-space line. However, this is less important for relatively high angles of elevation when large site-spacings are in use, since, for two paths in line with the satellite direction, the path extending furthest from the satellite will be well above any rain which obstructs the nearer path. Site diversity advantage will be substantially reduced if the heavy rain in frontal systems tends to affect both sites simultaneously, and this is most likely if the sites are aligned transverse to the prevailing wind.[180, 197] Conversely, convective cells tend to elongate along the prevailing wind and will produce an opposite effect on diversity gain.[198] In some regions radar measurements have shown the latter to dominate diversity performance.[65, 199] Local topographical features may also be important, so that two sites in one location may give markedly better diversity results than two sites similarly orientated a few tens of kilometres away.[152] This influence of microclimate should be noted in selection of sites for diversity operation. For a given site separation and direction to the satellite, the diversity gain has been shown to decrease as the elevation angle decreases, but the relationship with angle is not clearly established.[152] A cosecant law may be expected at low angles if attenuation due to widespread (stratified) humid clouds become significant. Having regard to present attenuation data collected on earth-space paths (see Section 5.3.1), site diversity is probably not cost-effective in mid-latitudes at frequencies less than about 12 GHz, but is worthwhile at 20 GHz and higher frequencies.

Whether or not spaced-site diversity is appropriate to a particular link, consideration should also be given to some other form of diversity system to reduce fading due to scintillation and multipath propagation, particularly for paths with low elevation angles (e.g. below $5°$) for which scintillation and multipath are most pronounced (see Section 5.3.2). In Section 4.6 it was suggested that the correlation coefficient for the two fluctuating signals of a diversity system should be less than 0·6, but in the limiting case of weak scintillation on two carrier frequency channels from a common antenna, this degree of decorrelation is achieved only if the carrier frequencies are separated by a factor of more than three.[200] *Frequency diversity* normally is not appropriate, both for this reason and because of the general preference of minimising use of extra frequencies. Emphasis is placed on weak scintillation because this is the situation least improved by diversity operation, and the system must operate over a wide range of scintillation conditions. Furthermore, although the isolation between orthogonal polarisations is usually high (see Section 5.5), *polarisation diversity* operation is also not of benefit because any fading on one channel is usually highly correlated with the fading on the other. However, *spaced-antenna diversity* is usually worthwhile where antenna costs are not too high, and an antenna separation greater than 300 m transverse to the viewing direction is recommended for use at 12 GHz.[201]

5.5 Use of orthogonal polarisations

In order to increase the available frequency capacity on earth-satellite links, it is useful to be able to transmit separate channels on two orthogonal linear or circular polarisations. However, the technique is limited by *cross-polarisation* (coupling) causing a reduction in cross-polarisation isolation (XPI) and cross-polar discrimination (XPD). As mentioned in Section 4.7 when defining these terms, many experimental measurements are of XPD rather than of XPI, but the two are numerically equal if degradation is due to rain. Cross-polarisation is produced by certain propagation factors, but even when these factors are not present, there will be some cross-polarisation due to the antennae systems themselves.[315] Many of the points discussed in Section 4.7 will be relevant here, although there are three marked differences. First, on terrestrial line-of-sight paths, it is possible to take advantage of the axes of the raindrops being nearly vertical by selecting vertical and horizontal polarisations so as to minimise cross-polarisation. The effect of the slight canting of the raindrops from vertical is small. However, for earth-satellite paths, the satellite orientation will generally preclude the use of true horizontal and vertical polarisations (as seen at the ground), and variations in the canting angle of raindrops then may produce marked changes in the XPD.[71]

The second, and rather obvious difference from terrestrial paths is that earth-satellite paths are seldom at low elevation angles. One consequence of this is that the multipath effects considered in Section 4.7 are not applicable, and another is that the reduction in XPD due to rain is less severe. As yet it has not been possible to produce a clear *relationship between cross-polar discrimination* D (dB), and co-polar attenuation A (dB) due to rain, but for a path elevation angle α° and linear polarisation in a plane inclined τ° from the vertical, it appears to have the form[317]

$$D = 30 \log f - 20 \log A - 40 \log (\cos \alpha) - 20 \log (\sin 2\tau) \qquad (5.4)$$

E.T.O.R.—K

where $10 \text{ dB} < D < 40 \text{ dB}$, $8 \text{ GHz} < f < 40 \text{ GHz}$, $1 \text{ dB} < A < 15 \text{ dB}$, $\alpha < 60°$ and $10° < \tau < 80°$.

For vertical or horizontal polarisation, when $\tau = 0$, the last term in eqn. 5.4 becomes $+ 12 \text{ dB}$. For circular polarisation the term involving τ disappears, i.e. the cross-polarisation is the same as for linear polarisation with a tilt angle $45°$ to the horizontal. For very low angles of evelation, $\cos \alpha \approx 1$ and eqn. 5.4 is reduced to eqn. 4.32.

A third major difference from the cross-polarisation effects on most line-of-sight paths is that *ice particles* well above the height of the melting layer may produce significant cross-polarisation effects on earth-satellite paths even when the attenuation is small. For this reason it has sometimes been referred to as 'anomalous depolarisation'. Because ice cloud and rain may both contribute to cross-polarisation, for many climates it is of rather limited value to seek a relationship between attenuation and cross-polarisation of the form of eqn. 5.4 for earth-satellite paths.

The relative importance of cross-polarisation due to rain or ice clouds is dependent on climate and frequency.[202] Although rain may cause the lowest values of XPD for small percentages of the time, in the 20/30 GHz band these values correspond to unacceptably large co-polar attenuations. Thus cross-polarisation interference due to ice clouds may be more important than rain effects in this band, despite the attenuation being relatively low. However, it seems likely that rain attenuation will remain the major cause of system failure due to propagation effects at these frequencies. For circularly-polarised systems operating in the 11/14 GHz band with fade margins in the 7–12 dB range, the effects of cross-polarisation interference and rain attenuation on failures are more nearly comparable. For linearly-polarised systems with polarisation tilt angles giving near maximum improvement over circular polarisation, attenuation will have a greater influence. Only in the 4/6 GHz band should cross-polarisation interference have a greater influence on failures than rain attenuation, and here perhaps only for circular polarisation, or for linear polarisation with least favourable tilt angles. Measurements indicate that little cross-polarisation is produced by the melting layer. Depolarization due to hydrometeors is of no practical importance at frequencies below 1 GHz.

Measurements of XPD are very limited, but such measurements as there are suggest that the XPD exceeded for 99·99% of time was 11 dB for a $9°$ elevation angle in Japan, and 24 dB for a $20°$ elevation angle in Taiwan, both for 4 GHz circularly-polarised transmissions, and about 20 dB for elevation angles of 20 to $25°$ in Western Europe for linearly-polarised waves at 20 and 30 GHz.[317] All these XPD values are above the minimum acceptable for systems employing frequency re-use, but results obtained on paths at low elevation angles would probably not be acceptable. For locations in which the co-channel interference is unacceptable, an adaptive cross-polarisation cancellation device may be used. Measurements have shown two effects that may cause difficulty in designing such cancellation systems. The first of these is that there is often a very rapid change in depolarisation at the instant of a lightning discharge.[203, 204] This suggests a change in the orientation of the ice crystals, a postulate which is supported by radar data.[205] The effect is important, since a failure of the cancellation system to follow a large and sudden change, may for an instant lead to a large error rate in digital signals. The second effect is that, although in general the relative phase of the co- and cross-polar signals is fairly constant (i.e. generally within $\pm 20°$), abrupt changes of about $180°$ can occur under certain conditions.[204, 205] This may produce a requirement for phase and amplitude cancellation.

Although it is not normally of consequence for earth-to-satellite links operated at frequencies above UHF, peak values of *ionospheric Faraday rotation*[315] may be a more significant cause of cross-polarisation than hydrometeors in some locations for SHF links using linear polarisation. The effect may be overcome by the use of circular polarisation. The rotation is inversely proportional to the square of the radio frequency, and it may vary by a factor of two or three depending on solar activity and latitude. Calculations have shown that as many as 10 rotations might be expected on a $30°$ elevation 1 GHz path in the worst conditions.[172]

5.6 Propagation delays

The mean and standard deviation of the signal time delay relative to free space, i.e. the range error, on an earth-space path through a precipitation-free troposphere can be derived from its empirical relationship with the surface refractive index.[2, 207] The delay is in part $\Delta \tau_N$, due to the reduced phase velocity in the refractive medium, and in part $\Delta \tau_c$, due to the difference between the curved ray path length and the true slant range. Both are related to the angle of elevation, and are given in Table 5.2 for a path extending throughout an atmosphere with an exponential model of refractive index decrease with height and a surface refractive index of 320 N units.[2] For larger values of surface refractive index, the time delays will be larger than those shown. Clearly, the delay due to ray curvature, compared with that due to reduced phase velocity, is significant only at very low elevation angles.

Table 5.2   *Propagation delays on a path through the troposphere due to reduced velocity $\Delta\tau_N$, and due to ray curvature $\Delta\tau_c$*

Elevation angle at ground	$\Delta\tau_N$	$\Delta\tau_c$
	ns	ns
0°	330	33
1°	220	9·7
3°	123	2·2
10°	42	0·083
30°	16	0·003

Based on data from Bean and Dutton[2]

Whereas at frequencies above 10 GHz time delays due to the ionosphere are negligible compared with those due to the troposphere, the converse is true at VHF and UHF. For VHF and higher frequencies, the signal delay through the ionosphere $\Delta\tau_i$(s), is given by

$$\Delta\tau_i \simeq 1{\cdot}3 \times 10^{-7} N_T / f^2 \tag{5.5}$$

where N_T (el/m^2) is the integral of the electron concentration along the ray path, and f(Hz) is the frequency.[312] By way of example, for an integrated electron content of 2×10^{18} el/m^2 (which is probably the maximum value that might occur for a 30° elevation angle), the group delay would be 26 μs for a frequency of 100 MHz but only 2·6 ns for a frequency of 10 GHz.

Besides knowing the average delay (or range error) in earth-space propagation caused by the atmosphere, it is important for both satellite ranging and for synchronisation in digital satellite communication systems also to know the variance of the propagation delay. Measured values of the standard deviation of the time delay depend on the time duration over which measurements are made. Means are available for determining the value that would occur over one time period from data obtained over another time period.[317] Over a two-way earth-space path inclined at 30° elevation, the standard deviation has been found to be about 0·02 ns when measured over a minute, about 0·2 ns over an hour and about 0·6 ns over a day. In each case it is proportional to the cosecant of the elevation angle. In addition to refractive index effects, rain may influence the fluctuations in propagation delay which occur over a period of a few minutes or so. However, the propagation delays due to rain and refractive index effects are both so short that they impose no practical limitations to bandwidths used on satellite communications systems, nor do they restrict the guard times used on time division multiple access systems.[207, 208]

Because the refractive index in the ionosphere is a function of frequency, its effect on group delay is particularly important in respect of the *dispersion* which it introduces. This dispersion is such that the differential delay $\delta(\Delta\tau_i)$ across a bandwidth δf is given by

$$\delta(\Delta\tau_i)/\delta f \simeq -2\cdot7 \times 10^{-7} N_T/f^3 \qquad (5.6)$$

By way of example, for an integrated electron concentration of 2×10^{18} el/m^2, as above, the dispersion would be $0\cdot5$ ps/Hz at a frequency of 100 MHz, but only $0\cdot004$ ps/Hz at 500 MHz.[172, 312] As was implied earlier there is no dispersion in the troposphere except at the edges of the main absorption bands above 50 GHz.

Chapter 6

Transhorizon paths

6.1 Introduction

Under conditions of standard refraction, the transmission loss at frequencies above 30 MHz increases reapidly beyond the horizon, the rate of loss with distance being determined by *diffraction* over the intervening terrain. When the transmission loss has increased some tens of decibels in excess of the free-space value, the rate of attenuation with further distance falls to a much smaller value, about 0·1 dB/km, since the field due to scatter from tropospheric irregularities is then much greater than the diffraction field. By this propagation mechanism reliable *'tropospheric scatter'* (or 'troposcatter') links may be established over large distances (up to 1000 km) for the transmission of small-capacity multiplex-telephony signals without relay stations, and over medium ranges (200–300 km) for the transmission of a large number of telephoney channels or of TV channels. However, signal fading is normally severe. Such links require high-power transmitters, high-gain antennae, sensitive receivers and (usually) some means of diversity operation to overcome signal fading. High-gain antennae require large reflecting surfaces, and the lower frequency limit to practical operation has been about 200 MHz. The upper frequency limit of about 5 GHz is caused by the increase in attenuation with frequency (as discussed in Section 6.2) and attenuation due to precipitation, which becomes appreciable on the long paths. For a small proportion of time abnormally high signal levels (i.e. anomalous propagation or 'anaprop') due to ducting or reflection of radio energy occur on 'troposcatter' links. These conditions do not relate to the reliability of such links except in so far as they may be responsible for interference between one radio link and another, a subject discussed in Section 8.3.

For radio paths extending only slightly over the horizon, or for certain paths extending over an obstacle or over mountainous terrain, diffraction will generally be the propagation mode determining the transmission loss. Because the rate of increase of transmission loss is so rapid for paths just beyond the horizon, it is not normally practical to set up such a point-to-point radio link. It is more likely that such a path would be used near the limits of the service area of a transmitter used for broadcasting or mobile radio, and this topic is discussed in Chapter 7. However, in exceptional cases a diffraction path over a pronounced mountain ridge may produce a reliable radio link if the ridge approximates to a knife edge (see Section 7.2). Since diffraction losses are normally large, metallic surfaces are sometimes suitably located to reflect energy to and from the remote terminal when it is essential to operate such a path without an active repeater.

In Section 2.7 it was emphasised that it is difficult to *predict* the magnitude of the forward-scattered power from a transmitter antenna beam to a receiver antenna beam. Not only is the variance of the refractive index itself variable with time and position, but so too is the wave number spectrum, and these factors change with climate, with local meteorological conditions and with height. As a result, it is not only very difficult to obtain a reliable estimate of forward scatter, but also to apply data obtained on one radio path to assess that to be expected on another for which the path length, aerial height, terrain type, antenna beams and radio frequency may all be different. Empirical relationships have been established from a large accumulation of data, but any particular path may have a very different median field and range of fading from that predicted.

However, some form of prediction is necessary, and established methods are considered in Sections 6.2 and 6.3. The slowly varying statistical distributions of transmission loss about the long-term median value are found to be largely independent of the median value itself. It is, therefore, expedient to consider first the long-term median value, and then the slow variation about this median. These determine whether a link will be operable with given path and terminal parameters. Superimposed on the slow variation of signal level there will be a rapidly varying fading pattern which determines what use can be made of such a link in terms of

(*a*) the transmissible bandwidth
(*b*) the associated limitations to rate of transfer of digital data over the link or
(*c*) the gain degradation of the antennae that may be expected.

Problems associated with rapid fading may be overcome to some extent by suitable siting of the terminal stations and by diversity operation.

6.2 Prediction of long-term median transmission loss

Although it is not normally possible to give a reliable prediction of the signal level to be expected on a given radio path, it has been possible to examine separately the factors most likely to influence the transmission loss, and then to form a semi-empirical expression summing the contributions from these various factors. This has led to a formula for the *annual median transmission loss* due to tropospheric scatter of the following form.[209, 308]

$$L(50) = 30 \log f - 20 \log d + F(\theta d) - (G_t + G_r - L_d)$$
$$- V(d_e) \qquad \text{dB} \qquad (6.1)$$

where the angular distance θ (radians) is the angle between the radio horizons in the great-circle plane containing the path terminals for median atmospheric conditions, f(MHz) is the frequency and d (km) the path length; L_d is the combined aperture-to-medium coupling loss (or gain degradation) of the two antennae of gain G_t and G_r (dB) and is given by the semi-empirical relationship

$$L_d = 0.07 \exp [0.055 (G_t + G_r)] \qquad \text{dB} \qquad (6.2)$$

for values of G_t and G_r each less than 50 dB, for distances between 150 and 500 km and for frequencies ranging from 400 to 10 000 MHz (see Section 6.4). The term $V(d_e)$, a function of the 'effective distance' d_e (defined in Section 6.3), is a correction for various types of climate, and varies between a maximum of 7·5 dB (for a maritime temperate climate at 250 km effective distance) to a minimum of −8·5 dB for paths over desert with effective distance greater than 350 km, or 600 km in equatorial climates. The function $F(\theta d)$ has a small dependence on surface refractive index N_s, as is shown in Fig. 6.1. In temperate regions, where the greatest number of observations have been made, monthly-median transmission losses tend to be higher in winter than in summer, with a difference of 8-10 dB between the mean values, but the seasonal effect diminishes as the distance increases. For equatorial climates the seasonal variation is small. There is negligible variation in the median value from year to year so that the terms annual median and long-term median are essentially synonymous.

Using the notation of eqn. 6.1, and comparing with eqns. 1.4 and 1.7, the long-term median path attenuation with respect to free-space is

$$A(50) = L(50) + G_p - L_{bf} \tag{6.3}$$

i.e.

$$A(50) = 10 \log f - 40 \log d + F(\theta d) - V(d_e) - 32 \cdot 4 \quad \text{dB} \tag{6.4}$$

The dependence of this attenuation on distance is governed by the relative influence of the terms $40 \log d$ and $F(\theta d)$ and this will change with the values of d and θ. If it is assumed that the antenna beams are

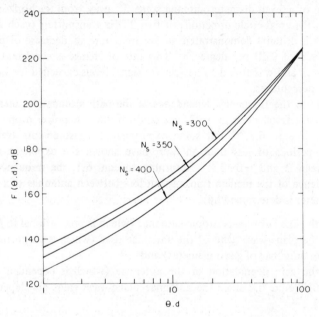

Fig. 6.1. *$F(\theta d)$ as a function of surface refractivity N_s*
[Based on CCIR[308]]

horizontal at the path terminals, i.e. $\theta = d/a_e$, then Fig. 6.1 shows that $F(\theta d) - 40 \log d$ remains almost constant at 52 dB for $200 < d < 1000$ km, i.e. the attenuation remains constant apart from the small variation in $V(d_e)$ mentioned above. However, if the elevation angles of the antenna beams above horizontal add to $1°$ or $2°$, then $F(\theta d) - 40 \log d$ increases uniformally by about $0 \cdot 05$ dB/km within the same range of distances.

The rapid increase in attenuation on a path [as contained in $F(\theta d)$] when the antenna beams are directed well above the horizontal occurs because

(*a*) the volume in which scattering between the beams occurs is at a height where the refractive index fluctuations that cause scatter are less intense, and

(*b*) the volume common to the antenna beams is reduced.

This effect is very marked, so that considerable care should be taken to ensure low and unobstructed horizons for the terminal points. The precise optimum antenna beam elevation is a function of the path and atmospheric conditions, but it lies within about 0·2 to 0·6 beamwidth above the horizon.[210] Measurements made by moving the 0·4° wide beam of a 25 m diameter antenna some 3° away from the horizon and from the great-circle directions of two 2 GHz transmitters (each about 300 km distant) demonstrated an apparent rate of decrease of power received of 9 dB per degree.[211] This rate of change is consistent with Fig. 6.1 and was found to be true for signal levels exceeded for various time percentages.

As to the frequency dependence of the path attenuation, measurements at frequencies up to 3 GHz confirm the first power proportionality of eqn. 6.3,[308] but some experimental measurements averaged over periods of less than an hour have shown the exponent to be between 2 and $-0·33$.[212] Returning to eqn. 6.1, the frequency dependence of the median transmission loss between antennae of a given diameter is determined by

(*a*) the loss between isotropic antennae (which is proportional to f^3)

(*b*) the plane-wave gains of the antennae used (which is proportinal to f^2 for antennae of given diameter) and

(*c*) the gain degradation of the antennae (which is dependent on f through eqn. 6.2 since the antennae plane-wave gains are dependent on f).

The effect of these three factors as a function of frequency for antenna diameters between 3 and 30 m is shown in Fig. 6.2.[313] This Figure represents the relative loss between the terminals of two antennae of the same diameter located at the ends of a transhorizon radio path, which is assumed to be between about 150 and 500 km in length. The reference loss (0 dB) is taken as that which exists under the same conditions between two antennae 10 m in diameter at 1000 MHz. The optimum operating frequency lies between 300 MHz for an antenna with a diameter of 30 m, and 3 GHz for an antenna of 3 m diameter. However, the practical frequency band for any antenna pair is wide since the loss minima of Fig. 6.2 are very flat. Although the antenna dimensions necessary to achieve a given antenna gain are smaller at higher frequencies, the mechanical tolerances are more severe, and so the maximum useful diameter of antennae is about 130 times the wavelength for frequencies above 1 GHz.[313] This corresponds to a plane-wave gain of about 50 dB and a gain degradation of about 15 dB for two identical antennae.

At frequencies greater than about 5 GHz, the additional attenuation due to absorption by atmospheric oxygen and water vapour is not

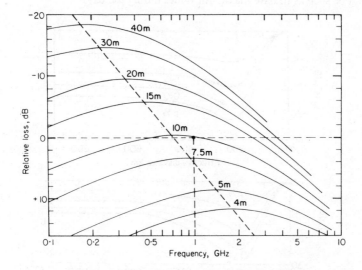

Fig. 6.2. *Relative loss between antenna pairs each of the diameter stated*
[Courtesy CCIR[313]]

negligible on long paths (see Section 3.4); and for small percentages of time attenuation by intense rain may also increase the transmission loss. However, should rain or ice clouds be particularly dense in the path common volume, the forward-scattered signal may (in principle) be increased, and so partially offset transmission loss.

6.3 Slow signal variations

Slow variations of transmission loss on a radio path may have cyclic periods of

(*a*) several hours
(*b*) one or more days or
(*c*) seasons.

They are caused by major changes of refractive index characteristics on the radio path and may be associated with the general weather type.

The slow fading characteristics are therefore a function of climate, but they are not strongly dependent on the radio frequency. Diurnal variations tend to produce maximum transmission loss in the afternoon, particularly during the summer, and on relatively short paths (100–150 km). At much greater distances, e.g. 600 km, the diurnal cycle can be such as to have maximum values during the day.

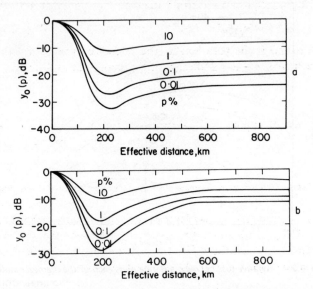

Fig. 6.3. *Variation with effective distance of the hourly-median transmission loss exceeded for time percentage p compared with median value*
a For an overland path in a maritime temperate climate (e.g. UK)
b For a continental temperate climate (e.g. Continental Europe and USA)

[Courtesy CCIR[308]]

A useful parameter for predicting the reliability of a transhorizon system is the *distribution of hourly median transmission loss*.[308] The hourly median values of transmission loss in excess of the long-term median are distributed approximately log-normally, with a standard deviation which generally lies between 4 and 8 dB, depending on the climate and path lengh. An indication of the transmission loss $L(p)$ exceeded for a certain time percentage p is given as a function of the 'effective distance' by Figs. 6.3 *a* and *b* for two of the nine climates considered by the CCIR, where

$$L(p) = L(50) - y_0(p)\, g(f) \tag{6.5}$$

The frequency-dependent parameter $g(f)$ is approximately 1·0 at both 200 MHz and 1 GHz, and has a maximum of about 1·15 at 400 MHz.[308]

The *effective distance* d_e (km), as used in eqn. 6.1 and Fig. 6.3, is arbitrarily defined for $d \leqslant d_{ref}$ by

$$d_e = 130\, d/d_{ref} \tag{6.6}$$

or, as is generally the case, for $d > d_{ref}$ by

$$d_e = 130 + d - d_{ref} \tag{6.7}$$

where d(km) is the path length. The reference distance d_{ref} (km) is shown in Fig. 6.4 to be the sum of the smooth-earth distances to the radio

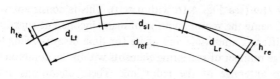

Fig. 6.4. *Definition of reference distance for evaluation of effective distance*

horizon d_{Lt} and d_{Lr} (km) (dependent on the effective antenna heights h_{te} and h_{re} (km) above the foreground terrain), and the distance d_{sL} (km) for which the diffracted and scatter-type signals are equal. For a radio frequency f(MHz), these are taken to be

$$d_{s1} = 300\, f^{-1/3} \tag{6.8}$$

and

$$d_{Lt} + d_{Lr} = (2a_e h_{te})^{1/2} + (2a_e h_{re})^{1/2} \tag{6.9}$$

and the effective earth radius a_e is assumed to be 9000 km.

It has been observed that the variability of the hourly transmission loss is usually greatest for path lengths d slightly greater than d_{ref}.[308] An estimate of the *standard error of prediction* for any given percentage of the time is given by

$$\sigma(p) = [13 + 0·12\, y_0^2(p)\, g^2(f)]^{1/2} \qquad \text{dB} \tag{6.10}$$

The use of equations 6.5 and 6.10 with those of Section 6.2 for estimating interference between systems at times when transmission losses are low is considered in Section 8.3.2.

In addition to statistics for the year, planning of point-to-point links is often based on the transmission loss likely to be exceeded for as little as 1% of the *worst month*. A method has been prepared which enables the transmission loss between isotropic antennae to be derived graphically, given the equivalent path distance (the product of the angular distance and the effective earth radius), the time percentage required, the type of climate and the frequency.[213, 308]

6.4 Rapid fading and its associated effects

The rapid fading which occurs on radio paths extending well beyond the horizon, when the received field-strength is due to troposcatter, has typically a frequency of a few fades per minute at VHF, 0·5 to 1 Hz at UHF and a few hertz at SHF. If the fading pattern is analysed over periods of up to 5 min the amplitude distribution generally follows a *Rayleigh law* (see Fig. 4.16 with $\sigma = 0$). This is consistent with the signal level being the vector addition of many incoherent contributions, i.e. with differing amplitude and phase (or time delay), owing to there being a large number of scattering elements within the common volume of the two antennae of the radio link. These conditions are to be expected for the median and low signal levels associated with scatter; but for some relatively high signal levels too, there may be rapid fading from contributions generated by the common volume containing a few highly reflecting layers of relatively small horizontal extent (sometimes called 'feuillets' or 'leaves'), since a distribution which is indistinguishable from a Rayleigh law also results from a relatively small number of components (e.g. 5) having random relative phase. For propagation conditions where the common volume contains only one or two strongly reflecting layers (see Section 2.6), the transmission loss may become very low; the short-term fading range small and the fading frequency low, but these last conditions relate more to paths likely to cause interference (see Section 8.3.2) than to reliability of radio links.

Fading frequency distributions have been studied in terms of the time auto-correlation function, which provides a *'mean fading frequency'*.[214] This mean fading frequency varies in a statistical manner between 0·044 (exceeded for 99% of the time) and 4·3 (exceeded for 1% of the time) times its median value. The median value itself increases nearly proportionately to distance and carrier frequency. It may reach 4 Hz with a carrier frequency of 10 GHz. When larger antenna diameters are used, this fading frequency decreases slightly due to a spatial averaging of the non-uniform wavefront (i.e. a restriction in the angular range of scatter contributions being received). Fading frequencies up to about 20 Hz have been observed during the influence of rainfall on a 210 km path operating at 12 GHz.[215]

Measurements made in the UK at 2 GHz on two paths of 300 km demonstrated the way in which the distribution of fade rate changed as the hourly median level itself changed.[211] With hourly-median levels of path attenuation exceeded for less than about 0·2% of hours, fades were not observed more frequently than 3 times per minute; but with attenuation greater than the long-term median value the range of fade rate was from 6 to 120 per minute. Similar measurements in the USA at 186 MHz on paths of 376 and 633 km showed several fades per minute at the long-term median level, several minutes per fade at the level exceeded for 5% of hours, and virtually no fading at the level exceeded for 1% of hours.[27] Other measurements of the *duration of fades* have shown that, over periods of 1 to 5 min, the field strength obeys Rayleigh's distribution law sufficiently closely to render relatively easy the calculation of the effect of diversity reception.[216, 217] The phase of the received field also varies randomly, with a constant density distribution. Sudden deep and rapid fading has been noted when a frontal disturbance passes over a link. Also, reflections from aircraft can give pronounced rapid fading.

Various experimental studies have been made of the *angular distribution* of incident energy on troposcatter links at frequencies above 3 GHz.[211, 218, 219, 320] This is important when considering the range of elementary paths contributing to gain degradation and limiting transmission bandwidth. From such studies it is possible to distringuish times when the incoming signals arrive over a very narrow range of angles, the signal level is high and the fading is very small and of small amplitude. At other times, when the signal level is less high and the fading more pronounced, the signal level has been found to be proportional to factors ranging between $\Delta\theta^{-4}$ and $\Delta\theta^{-10}$, where $\Delta\theta$ is the observed range of angles of incident energy in both elevation and bearing.

One consequence of having a large angular spread of radiation incident on an antenna is that greater phase incoherence occurs across the aperture of a larger antenna than a smaller one. From an alternative viewpoint, the angular spread of incident radiation may be encompassed by the wide acceptance beam of a small antenna, but only a relatively small proportion of the incident radiation may be accompanied by the narrow acceptance beam of the large antenna. The consequence is that the narrow-beam antenna does not realise its planewave gain, compared with an isotropic antenna, i.e. there is an *aperture-to-medium coupling loss* (or gain degradation) L_d, for the larger antenna.[221, 222] Evidently the number of paths through the atmosphere over which energy may be carried from the transmitter to the receiver depends on the gains (i.e. the beamwidths) of both antennae, and the magnitude of the aperture-to-medium coupling loss on a troposcatter path for median signal levels, as given by eqn. 6.2. This is in agreement with experimental observations to within 1 dB for $G_t + G_r < 100$ dB.[223, 224]

As the sum of the plane-wave gains $(G_t + G_r)$ increases, the rate of increase of the path antenna gain $(G_t + G_r - L_d)$ is less rapid, and for very large gain antennae the rate of increase of the path antenna gain is found to be only 25% that of the plane-wave gain,[225] the gains being expressed in decibels. This has considerable practical importance, since to increase the useful gain by 1 dB it is necessary to increase the plane-wave gain by 4 dB, which means enlarging the diameter by about 60%. Indeed, one may conclude that building antennae with plane-wave gains in excess of 45 or 50 dB would give little signal benefit for most troposcatter paths in European-type climates, though there would still be an advantage in acquiring more available bandwidth. Apart from the median aperture-to-medium coupling loss, it is also important to know the value that occurs for low signal levels, e.g. for signal levels exceeded for 99% of the time, when high antenna gain is most needed. The aperture-to-medium coupling loss for each antenna may then be 2 or 3 dB greater than the median.[223] If, however, the angular distribution of the incident energy is much narrower, e.g. some form of ducting causes the incident wave to be closer to a plane wave, then the aperture-to-medium coupling loss may also be much less. This is very relevant to interference from or to tropospheric scatter link terminals (see Section 8.3.2).

Mention has been made that because scattered contributions to the total energy received travel over different paths and arrive from a great number of different directions with a range of path delay times, there is a limit to the *available bandwidth*. If Δt(s) is the difference in path delay time between the shortest and the longest paths, this bandwidth B(Hz) is defined by eqn. 4.29. If the antenna has a broad beam, the longest delays, and hence the bandwidth, are determined by the propagation medium; but it is possible to reduce the longest path delays, thus increasing the bandwidth, by using high-gain narrow-beam antennae. Fig. 6.5 shows the way in which the available bandwidth increases with the free-space antenna gain and decreases with the path length. It is based on a study made at UHF,[226] and this approach has subsequently been extended.[227] The distance dependence occurs because, for a given antenna beamwidth, the maximum delay will be larger for longer paths. Because the off-great-circle spreading of energy depends on weather conditions, the statistics of transmissible

bandwidth depends on the climatic zone. Experimental studies show
that an increase in the antenna gain (which reduces the angular range
over which ray paths may exist), widens the transmissible bandwidth
to the extent to which the aperture-to-medium coupling loss also
increases (for gains exceeding about 30 dB).[225]

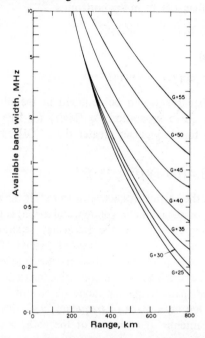

Fig. 6.5. *Frequency bandwidth that may be used on a tropospheric scatter
path, as a function of range and antenna gain in free space*
[Courtesy Picquenard[124]]

6.5 Use of diversity

To some extent the various effects associated with rapid fading can be
overcome by diversity operation, and to achieve an operational reliability of about 99·9% of time over a tropospheric scatter path, as is
required for high grade communication channels, may require operation
in two modes together, i.e. quadruple diversity. The most widely used
system at present is a combination of *space and frequency diversity*.
Such a system, and the improvement that may be obtained for a
Rayleigh fading distribution, was considered in general terms in Section
4.6. The antenna spacing required for space diversity operation has
been studied by measuring the spatial correlation function and its

scale length l from simultaneous recordings of the transmission loss with horizontally and vertically spaced antennae. For paraboloidal antennae, an adequate separation $\Delta(m)$, depends on $l(m)$ and on the diameter of the antennae D(m). Using the respective values of l exceeded for 1% of time in the horizontal and in the vertical, then, for frequencies greater than 1 GHz, in the horizontal[308]

$$\Delta_h = 0.36 (D^2 + 1600)^{1/2} \tag{6.11}$$

and in the vertical

$$\Delta_v = 0.36 (D^2 + 225)^{1/2} \tag{6.12}$$

Much greater distances would be required in the direction of propagation. An adequate separation Δ_f (MHz) for frequency diversity operation, again for frequencies greater than 1 GHz, has been found to be[308]

$$\Delta_f = (1.44 f/\theta\, d)(D^2 + 225)^{1/2} \tag{6.13}$$

where f(MHz) is the carrier frequency, θ (mrad) is the angle of scattering in the centre of the common volume, and d(km) is the path length. A general working rule is that the frequency separation should be about 1%, and the antenna spacing should be about 100 wavelengths, but on shorter paths (e.g. about 200 km) the frequency separation should be increased to 3% and the antenna spacing to 200 wavelengths.

As well as the space/frequency mode of quadruple-diversity operation, some use has been made of *frequency/angle diversity*. This employs only one antenna at each end of the path, but two vertically spaced feeds to the receiver antenna so as to give offset beams. This technique is useful where large antennae and high land costs are involved, or where space is not available for a second antenna, as for example on an offshore oil rig platform. Vertically-spaced antenna beams are preferable to horizontally-spaced beams because the refractive index irregularities change character more rapidly with height than with horizontal position (the same point would favour vertically-spaced antennae for space diversity, but the required separation is usually somewhat more difficult to achieve). A beam separation of 6 mrad has been found to produce a correlation coefficient of 0.4 at a frequency of 2 GHz with 10 mrad beamwidth antennae on a 250 km long path.[228] Some allowance has to be made for the fact that the signal level received on the upper beam tends to be less than that received on the lower beam, and there is also a small 'squint loss' due to each of the two antenna feeds each being offset from the antenna prime focus.[229] However, the method has the advantage of being the only diversity system using completely separate common volumes, which gives a high probability of obtaining a low long-term correlation coefficient between the channels. A simple angle-diversity system at the receiver end of a link may provide a performance comparable to (and considerably more economical than) that of a space-diversity system, particularly in the SHF range.[229, 230] The performance is also comparable with that of frequency diversity.

In general, for real time transmission, space and/or angle diversity are beneficial so far as spectrum conservation is concerned. However, *time diversity* (usually with the addition of error correcting codes) may also be of interest when real-time transmission is not a requirement, and where space and spectrum considerations are important. This is possible on transhorizon scatter links, even when the signal level is occasionally fading into the system noise level, since the durations of fades are fairly brief, and so the data holding times are fairly short.

Chapter 7

Area coverage

7.1 Introduction

The last three chapters have been concerned with reliability of point-to-point links, although when considering satellite-to-earth links (Chapter 5) it should be recognised that a 'service area' approach is becoming increasingly appropriate in respect of proposals for direct broadcasting from satellites, and direct contact with satellites in the maritime, aeronautical and land mobile services. In point-to-point services some choice may be exercised in selection of the link terminal sites to ensure optimum performance, and careful calculations of expected transmission loss may be made for specific paths. By contrast, this chapter is concerned with the coverage of a large area around a fixed terrestrial transmitter for the provision of a service such as sound or television broadcasting, or communication with mobile stations, where the remote terminal locations will normally be less than the ideal for radio signal reliability, and no specific path-profile information can be obtained. The area will lie both within line-of-sight and just beyond the horizon (or behind obstructions) and many of the general points made in Chapters 4 and 6 will apply.

Although broadcasting and land-mobile radio are somewhat different in their objective, the radio propagation problems associated with the two services are in many ways very similar. In each case the prime terminal, i.e. the broadcasting transmitter or communications base-station, can be selected, and the general coverage area can be predicted with some certainty. In the case of broadcasting, the transmitter antenna will normally be a considerable height above surrounding terrain, so as to cover as much land area as possible. However, there will be constraints on the antenna height for a mobile base station, since it must be high enough to give the necessary coverage area, but not high enough to give signal levels outside this area which will cause interference to other users of the same frequency elsewhere. A more particular difference between the two services is that a broadcasting receiver antenna can usually be sited to best advantage, whereas a mobile antenna passes through a radio field-strength pattern having a wide dynamic range. For this reason the fading problems of mobile radio communication are considered separately in Section 7.4. Most of the other material of the Chapter is applicable to both broadcasting and mobile radio, although sometimes the emphasis may be on one service or the other. Apart from fixed domestic receiver antennae, the remote terminal antenna may be on a vehicle or on a man-portable set for either broadcasting or land-mobile radio.

Maritime and aeronautical mobile services are not considered in detail here. The former is mainly carried out at frequencies below 30 MHz, where ionospheric effects are of importance rather than tropospheric effects. For the aeronautical mobile radio service, long-term-median prediction curves have been prepared for aircraft-to-ground and aircraft-to-aircraft use at frequencies from 125 to 15500 MHz, and for ranges up to 1800 km.[304] For line-of-sight conditions, the losses are assumed to be essentially those of free space, and beyond line-of-sight the methods of computation of the curves are essentially those used for land-mobile purposes (see Section 7.3.2).

The coverage areas for the broadcasting and mobile radio communication services are mainly line-of-sight (except for local obstructions), although they also extend for a short distance beyond the horizon. Both within and beyond line-of-sight the transmission loss is determined by diffraction, and this is discussed in Section 7.2. However, in most cases the roughness of the terrain and presence of obstacles do not allow diffraction theory to be applied to a simplified model. For this reason prediction procedures have been developed based on

(*a*) computer modelling for specific areas (see Section 7.3.1)
(*b*) graphical presentations for general areas (see Section 7.3.2)
(*c*) a simplified formula for quick estimations (see Section 7.3.3).

When considering the appropriateness of the VHF, UHF and SHF bands for either broadcasting or mobile services, a distinction may have to be made between providing a moderate service over as large an area as possible, or providing more complete coverage within a smaller general area. Specifically in the mobile context, where antenna location cannot be selected, long range may be more important for one user, and good local coverage for another. In either case, high transmitter antenna sites are desirable. From the various considerations in this Chapter the use of VHF is preferable to UHF if maximum range is the first consideration. However, if the prime objective is to provide an optimum service within a small coverage area, UHF is probably preferable. This is particularly true in city streets where the increased scattering from building surfaces will leave fewer 'holes' in the field-strength area coverage distribution. Particular consideration is given in this Chapter to reception in urban and suburban areas, as they are the areas of high population density where reliable serivces are required. Unfortunately, it is particularly difficult to predict signal levels in such areas, since there are so many variables.

In the mobile radio services, most uses are for voice communications rather than for data, although data-type signals may increasingly be carried with rates up to 1 or 2 kbit/s, i.e. using bandwidths comparable with speech. Experienced users of mobile radio, i.e. those in taxi, delivery, repair, fire, police or ambulance services, can manage with a signal/noise ratio of only about 10 dB, but for an inexperienced user operating a radio telephone, a higher signal/noise ratio is necessary. Like all mobile applications, radio telephone requirements are increasing rapidly, but it is for the broadcasting services that the highest quality of service is required, and it is those concerned with broadcasting who have taken the lead in producing good predictions of service coverage areas.

7.2 Diffraction over various terrain types

Some mention of diffraction has been made in Section 4.2.1 when considering diffractive fading due to obstacles close to a line-of-sight path. It is also the process most likely to determine the limit of service area coverage for braodcasting or mobile radio.

If a wave front is partially obstructed by an obstacle, some energy will be diffracted into the shadow region of the obstacle. In real situations diffraction effects will be complicated by the shape of the obstacle and by the effects of the atmosphere, but in many cases a simplified model can produce useful approximations to the effect obstacles have on radio links. For the case of a radio link with a very pronounced ridge of hills running across the line between a transmitter and receiver, this idealised model, known as *'knife edge' diffraction,* is a well established means of calculating diffraction losses. The diffraction loss A (dB), relative to the free-space loss, is shown in Fig. 7.1 as a function of the diffraction parameter v mentioned in Section 4.2.1. However, v is taken here as $+\sqrt{2n}$, where $\sqrt{n}$ is the height of the obstructing edge h_e normalised in terms of the radius r_1 of the first Fresnel zone at distance d_1 along the path, see Fig. 7.2. The magnitude of r_1 is given by eqn. 4.1 with $n = 1$, but now

$$v = h_e(2d/\lambda d_1 d_2)^{1/2} \tag{7.1}$$

where

$$h_e = \sin \theta \, (d_1 d_2/d) \tag{7.2}$$

where d, d_1 and d_2 are the distances indicated in Fig. 7.2, and λ is the

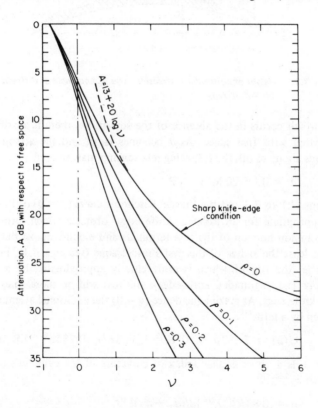

Fig. 7.1.   *Attenuation due to a 'sharp knife-edge' compared with a 'rounded knife-edge' (of reflection coefficient $R = -1$, see Section 4.2.3)*

v is a function of the diffraction angle (see eqns. 7.1 and 7.2)

ρ is a function of the radius of curvature of the edge (see eqn. 7.5)

Note the 6 dB loss for a sharp knife-edge ($\rho = 0$) at the limiting condition $v = 0$

[Based on CCIR[322] and Dougherty and Maloney[233]]

radio wavelength. All distances are in the same units, and θ is in radians. If the ridge is just below the line joining the terminals, then h_e, θ and ν are negative, a condition considered in Section 4.2.1. At the point where the screen just touches the direct line-of-sight, the field is half

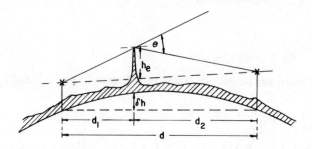

Fig. 7.2. *Height of knife-edge diffraction edge* h_e *and angle of diffraction* θ
$\delta h \simeq d_1 d_2 / 2a_e$

that which occurs in the absence of the screen, i.e. there is a 6 dB loss compared with free space. As θ becomes large, and the attenuation exceeds about 15 dB, the following relationship may be used

$$A = 13 + 20 \log \nu \quad \text{dB} \tag{7.3}$$

Eqns. 7.1 to 7.3 and the Fresnel-Kirchhoff curve ($\rho = 0$) of Fig. 7.1 are appropriate for an isolated knife edge obstacle which forms the the common horizon of the two terminals and extends across the path for at least the radius of the first Fresnel zone (see eqn. 4.1). For an obstacle, the top of which is more closely approximated by a wide *cylinder* (i.e. a rounded knife-edge), the loss will be more than that for a knife edge. At grazing incidence ($\nu = 0$) the additional attenuation is given by a term[231]

$$G(\rho) = 7 \cdot 192\rho - 2 \cdot 018\rho^2 + 3 \cdot 63\rho^3 - 0 \cdot 754\rho^4 \quad \text{dB} \tag{7.4}$$

where ρ is a dimensionless index of curvature of the cylinder's radius R given by[232, 233]

$$\rho = 0 \cdot 83 R^{1/3} \lambda^{1/6} [d/(d_1 + d_2)]^{1/2} \tag{7.5}$$

where all distances are in the same units. For regions well within the diffraction region, an additional term $E(\chi)$ must be added for propagation loss along the surface between the horizons,[320] so that for $-0 \cdot 971\rho \leqslant \chi \leqslant 0$

$$E(\chi) = G(\rho) \chi/\rho \tag{7.6}$$

for $0 < \chi < 4$

$$E(\chi) = 12\chi \qquad (7.7)$$

and for $\chi \geqslant 4$

$$E(\chi) = 17 \cdot 1\chi - 6 \cdot 2 - 20 \log \chi \qquad (7.8)$$

where

$$\chi = 1 \cdot 46\theta(R/\lambda)^{1/3} = 1 \cdot 25\rho\chi \qquad (7.9)$$

Examples of the attenuation for different hill top curvature is given in Fig. 7.1.

When diffraction losses become very large, radio fields produced by reflection from atmospheric elevated layers or by forward-scatter are likely to dominate (see Chapters 6 and 8). However, these propagation mechanisms are not likely to influence the service area of a transmitter in other than exceptional circumstances. More likely to be important is the fact that the height h_e in Fig. 7.2 and eqns. 7.1 and 7.2 is partly due to a component $d_1 d_2 / 2a_e$ which varies with the effective earth radius a_e. Allowance should be made for this variation in calculations on all but the shortest paths.

When a thin obstacle is surrounded by two plains with good surface reflectivity, there may be some modification to the path attenuation due to a *height-gain* effect caused by there being four paths for the waves to travel from the source to the receiver, as indicated in Fig. 7.3. Of the four paths, one is without reflection by the ground, two have one ground reflection and one path has two reflections. The height-gain pattern will therefore have distinct phase cancellation at certain localised heights, dependent on the geometry and wavelength. Since the phase conditions are critical, changes in effective hill height due to changes in effective earth radius due in turn to changes in refractive index gradient, may produce significant fading.

Fig. 7.3. *Diffraction and ground reflection for a knife-edge above a plain*

The theory of diffraction has been extended to propagation over two *successive knife edges*, and a method is available for predicting field strengths.[234] The theoretical solution for diffraction over multiple successive knife edges (three or more) requires the calculation of a multiple Fresnel-type integral of a dimension equal to the number of edges. For practical purposes, however, an approximate method is available which is based on a semi-empirical extension of single-edge diffraction.[235] This method consists of first calculating which obstacle would alone produce the greatest diffraction loss, then joining the summit of this obstacle to the antenna locations, and calculating the additional attenuation caused by the remaining obstacles according to the height they are above the lines so drawn, as indicated in Fig. 7.4.

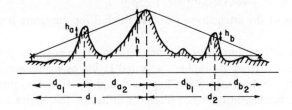

Fig. 7.4. *Multiple knife-edge diffraction*

Account should be taken of hill-top rounding. It is important to note from Figs. 4.2 and 7.1 that even if h_a or h_b (in Fig. 7.4) are slightly negative, those obstacles may still produce a small attenuation. No rigorous method exists for the prediction of field strength by diffraction over several rounded obstacles, but an approximate method is available which employs a simplified solution based on the assumption that each obstacle can be represented by a cylinder with a radius equal to that at the obstacle top.[236]

For a path on which there are no major diffraction edges, one may assume the losses to be those of diffraction over an assumed *smooth spherical earth*. The diffraction loss relative to free space A is then given by the sum of a number of terms[209, 320]

$$A = G(\chi_0) - F(\chi_1) - F(\chi_2) - 20.5 \qquad \text{dB} \qquad (7.10)$$

where $G(\chi)$ is associated with the whole path length and the terms $F(\chi)$ are associated with the two ends of the path. $G(\chi)$ and $F(\chi)$ are given in Figure 7.5, where

$$\chi_0 = d\,670(f/a_e^2)^{1/3} \qquad (7.11)$$

$$\chi_1 = (2a_e h_1)^{1/2}\,670(f/a_e^2)^{1/3} \qquad (7.12)$$

$$\chi_2 = (2a_e h_2)^{1/2}\,670(f/a_e^2)^{1/3} \qquad (7.13)$$

and where d (km) is the length of the radio path, h_1 and h_2 (km) are the antenna heights, a_e is the effective earth radius (km) and f (MHz) is the radio frequency. If the effective earth radius a_e is not constant along the path, account should be taken of this by using the appropriate values to calculate χ_1 and χ_2 (if a value for the maximum diffraction loss is required, i.e. the least favourable service condition, the minimum effective value of earth-radius factor along the path

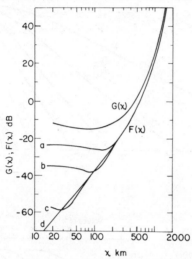

Fig. 7.5. *Diffraction formula terms $G(\chi)$ and $F(\chi)$ for the following conditions*
a Vertical polarisation over sea at 30 MHz
b Vertical polarisation over sea at 120 MHz
c Vertical polarisation over sea at 3300 MHz
d {Vertical polarisation over land above 600 MHz
{Horizontal polarisation over sea, or land above 100 MHz

[Based on CCIR[320]]

may be obtained from Fig. 4.4). For vertical polarisation, $F(\chi)$ is a function of frequency and ground type. As an alternative to using eqn. 7.10 to determine the attenuation, nomograms have been produced to achieve the same objective.[133, 320] The attenuation is expressed in terms of the loss over the path length around the earth's surface, less height-gain factors for the antenna height at each end of the path. Examples of the losses due to diffraction around a smooth earth surface are shown in Figs. 7.6 *a* and *b* for terminal heights of 100 m and 15 m.

Most often the radio paths of interest are over *irregular terrain* that has no pronounced features, sometimes called rolling terrain, for which neither the smooth-earth nor multiple-diffraction methods are applicable. For these conditions if has been found extremely difficult to calculate the effect of the roughness and irregularity of terrain features and environmental clutter, such as vegetation, buildings, bridges and electric power lines, except in terms of an empirically

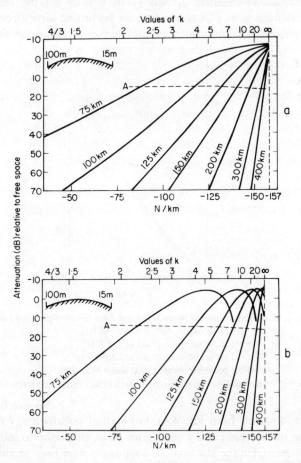

Fig. 7.6. *Attenuation due to spherical diffraction as a function of refractive index gradient*
  a For 4 GHz
  b For 12 GHz
  A: limit of visibility

[Courtesy CCIR[319]]

determined terrain factor. Consequently, it has been found practical to produce prediction curves for median conditions, and then to apply various empirical correction terms (as considered in Section 7.3.2), or a simple formula with an empirical terrain clutter factor (as considered in Section 7.3.3). In addition, theoretical diffraction methods have been developed to deal with certain models of terrain features[237, 238] and a fundamental theoretical approach to the problem of wave propagation over irregular terrain has also been made using an integral equation technique.[239] Some paths can be treated as a succession of crests, which may be represented as knife-edges or cylinders according to the sharpness of the crests.

A final point to be included in this Section is that of a pronounced obstacle visible from both terminals of a diffraction path producing '*obstacle gain*'. This gain is the difference between the attenuation over a radio path having such an obstacle, and the greater attenuation that there would have been in the absence of any such obstacle. It may occur when an otherwise multiple-diffraction path, or smooth-earth path, with high loss is transformed into a single-edge-diffraction path with less loss.[240, 241] Alternatively, an obstacle in the shadow of a diffracting edge may, by double diffraction, cause the received power level in its shadow to be as much as 15 dB greater than if it were not there. This has been put forward on theoretical grounds[234] and supported by laboratory scale-modelling with a laser beam.[242] A semi-empirical method for estimating such gain has been put forward and an atlas of the gains has been produced for certain obtstacles, antenna heights and frequencies.[243, 244] Also, pairs of neighbouring paths of comparable distances and antenna heights have been studied, one having a mountain ridge causing knife-edge diffraction, and the other being clear of obstacles. In this way the practical reality of obstacle gain has been established.[240, 241, 245-247] Although it is not normally likely to be a means of increasing significantly the power received on a transhorizon path, it may sometimes be that moving to a site behind a sharp topped ridge will produce an enhanced signal for a mobile terminal.

7.3 Service area coverage predictions

7.3.1 Computer methods

One means of expressing service area prediction is by maps bearing controus of the *probability of locations receiving an acceptable service* for at least p_T% of time, where this acceptable service must be expressed in terms of a minimum field-strength value.[306] Fig. 7.7 shows an example of such a map, which includes the effect of local hills and hollows and of the polar diagram of the transmitter antenna. In this case $p_T = 90$%. Such maps may be prepared from a large number of spot location measurements made over the area. As an alternative to direct measurements, use may be made of data derived from conventional survey maps and stored in computer-compatible *terrain data banks* in such a way that full point-to-point transmission loss calculations may be performed for any distance and radial direction from a

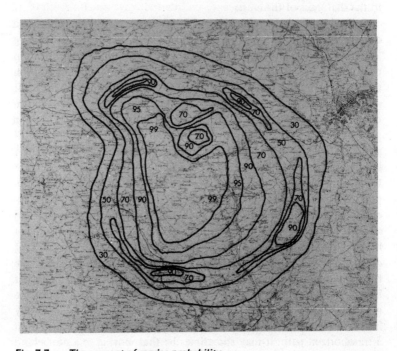

Fig. 7.7.   *The concept of service probability*
The numbers indicate the probability of locations receiving an acceptable service for at least 90% of the time

[Based on CCIR[306]]

transmitter.[248, 249] Limitations to basing the method on data from conventional maps are that

(*a*) heights of obstacles are not normally given
(*b*) the percentages of area covered by buildings are not readily derived
(*c*) the extent and character of woodland is not sufficiently clear.

However, propagation formulae have been optimised by comparison with measurements so that median field-strengths within a 50 m square at a height of 10 m may be predicted with RMS error of only 4 dB. Statistical models for the typical height of trees and buildings may be applied.

Fig. 7.7 gives a fairly detailed indication of the service area, but usually it is adequate to indicate with a single contour the area for which the field-strength is above a critical level for at least $p_T\%$ of time and for at least $p_L\%$ of locations using antennae of a given type. Typical values for VHF broadcasting might be 70 dB ($\mu v/m$) for 95% of time at 70% of locations. In practice, such maps are not fully indicative of broadcasting service areas because the user will give much more attention to the siting of a domestic receiver antenna in areas where the service probability is low than where it is high; hence the rather low figure of 70% of locations mentioned above. However, even if only a single contour level is used, then, provided good prediction methods are used with comprehensive terrain data banks, such maps do indicate where 'fill-in' transmissions may be required for either broadcasting or mobile radio application.

Apart from computer-based methods using detailed terrain data banks, simpler computer methods are also available which may be satisfactory in many applications, although these too need some topographical information,[250, 251] e.g. to assess the terrain irregularity. Such computer-based methods form the basis of those employing a terrain data bank and of those used to produce field-strength prediction curves (as described in the next Section). Good agreement has been obtained between measured and predicted data for relatively smooth terrain in rural areas, and these methods have since been extended for urban areas.[252, 253]

7.3.2 Field strength prediction curves

Where detailed data on the terrain characteristics are not available, it is useful to make use of prediction curves of *median signal-strength* as a function of distance, antenna heights, terrain type, frequency etc. Because those concerned with broadcasting have taken the lead in providing prediction methods, these have tended to be produced for high transmitter sites and the signal strengths have been measured 10 m above ground. This is a typical height for a roof-top antenna used for domestic reception. As a result of the large quantity of data now collected from different countries, fairly detailed prediction curves have been prepared for 30–250 MHz and 450–1000 MHz in terms of the field strength to be expected at distances of 10 to 1000 km from a transmitter.[302] Data have been given for the field strength exceeded for 50%, 10% and 1% of time and for 50% of receiver locations, for various transmitter antenna heights from 37 to 1200 m, for land, North Sea and Mediterranean-type conditions, and for a receiving antenna height of 10 m. From these, similar curves have been produced primarily for mobile applications for use with antennae at 1·5 m height in urban areas at 450 MHz and 900 MHz, and with 3 m antenna heights in rural conditions at 150 MHz.[318] In addition to data for the overland and over-sea paths, some curves are presented for mixed land-and-sea paths in terms of the relative percentages over sea and land.

Examples of curves for overland paths are given in Fig. 7.8. They apply for the rolling terrain found in many parts of Europe and North

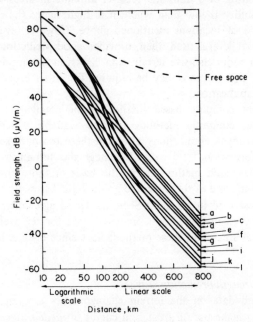

Fig. 7.8.   *Field strength decrease with distance from a 1 kW ERP half-wave dipole transmitter on overland paths*
Values exceeded for 1% and 50% of time for VHF bands for various transmitter heights. Free space values are given by eqn. 1.11 with $P_t = 0$ and $G_t = 2·2$ dB

Key to curves				
Frequency	Time		h_1, m	
	%	600	150	37·5
30–250 MHz	1	a	b	c
450–1000 MHz	1	d	e	f
30–250 MHz	50	g	h	i
450–1000 MHz	50	j	k	l

In each case $h_2 = 10$ m, $\Delta h = 50$ m and the values will be exceeded for 50% of locations

[Based on CCIR[302]]

America, for which the *terrain irregularity parameter* Δh is about 50 m. This parameter is defined as the interdecile range of the heights of the land above some datum level. The curves are multi purpose; as well as indicating radial limits to the service area under median conditions, they indicate signal levels to be expected on links extending well beyond the horizon (see Chapter 6) and levels of interference that may be exceeded for 1% of time at greater ranges (see Chapter 8). Particular points that may be seen from Fig. 7.8 for the shorter ranges (i.e. less than 20 km) are that the height depencence and distance dependence, approximately h^2 and d^{-4}, are consistent with eqn. 4.16. At longer ranges (i.e. greater than 400 km) the height dependence is approximately $h^{0.4}$, the distance dependence is approximately 0·1 dB/km, the UHF signal strength is less than that for VHF by about 10 dB at the median time level (8 dB at the 1% level) and the VHF signal level exceeded for 50% of time is 16 dB less than that exceeded for 1% of time (18 dB for UHF). More precise relationships may be found in Sections 6.2 and 6.3.

Various corrections may be applied to the curves for conditions other than those stated above. One of the largest of these corrections is to obtain the field strength exceeded for *percentages of locations* other than 50%. Measurements[131] show that the distribution with location is Gaussian, having a standard deviation σ_L of 8 dB at VHF and 10 dB at UHF if $\Delta h = 50$ m (18 dB at UHF if $\Delta h = 300$ m). This implies that, for 1% of locations, the received signal level will be a further 18 dB lower at VHF and 22 dB at UHF if $\Delta h = 50$ m (37 dB at UHF if $\Delta h = 300$ m). It is to be expected that σ_L increases with Δh, and that terrain irregularities have more effect at UHF than VHF, since the diffraction losses when the receiver is least favourably placed are an inverse function of wavelength (see eqns. 7.1 and 7.2). By normalising the terrain irregularity parameter Δh with respect to wavelength, the following relationship was found to give a good estimate of the location variability for a large number of measurements made in the USA for $\Delta h/\lambda$ less than 5000[254, 309]

$$\sigma_L = 6 + 0.69 \, (\Delta h/\lambda)^{1/2} - 0.0063 \, (\Delta h/\lambda) \tag{7.14}$$

It is found that the magnitude of the terrain irregularities affect not only the spread of transmission loss values with percentage of locations, but also the median loss. Again the irregularities are more important at UHF, where Δh of 150 m produces additional attenuation of 10 dB (7 dB at VHF) and Δh of 10 m produces a *reduction* of attenuation of the same magnitude.[302] For Δh of 300 m, i.e. very hilly or mountainous terrain, the losses are 10 dB at VHF and 15 dB at UHF (all these values for a path length less than 100 km).

Corrections may also be applied to the curves to obtain field strength values exceeded for different *percentages of time*. At VHF the standard deviation for both land and sea is about 3 dB for paths of 50 km, rising to 9 dB for paths of 150 km, and for UHF the values are 2 and 7 dB, respectively, over land, and 9 and 20 dB over sea.[318] The corrections for both different percentages of time and different percentages of locations refer to rural areas and not urban areas. It is significant for both broadcasting and mobile services that the standard deviation σ_{LT} of the field strengths sampled over a wide range of locations (with standard deviation σ_L) at a given distance from a transmitter and over a long period of time (with standard deviation σ_T) is given by[318]

$$\sigma_{LT} = (\sigma_L^2 + \sigma_T^2)^{1/2} \qquad (7.15)$$

In many cases it is not of interest to separate the effect of time and location, but to treat them as one distribution.

A major factor affecting the mean field strength is the *degree of urbanisation* surrounding the receiver point (assuming the transmitter to be sited well above influences of local terrain features). In suburban areas where the rooftop levels are fairly uniform, the received field strength is substantially the same as in rural areas so long as the antenna heights are well above local roof level. For lower antennae, the presence and density of trees may have an effect equal to that of the buildings themselves. At VHF and UHF the reduction of singal strength due to trees and buildings near to a broadcast receiver is typically 10 dB and 8 dB, respectively. In the centres of cities where very high buildings occur, the situation is more complicated and the shielding due to individual buildings may be as much as 30 dB.[255] The signal strength at any point may be made up from

(a) reflections off buildings
(b) diffraction around buildings and
(c) occasionally a direct line-of-sight to the transmitter with ground reflection present.

To some extent large diffraction losses around buildings may be compensated by scatter off other buildings. At 200 MHz it has been found that the transmission loss for a 10 m antenna height may increase by between 6 and 16 dB within heavily built-up areas, dependent on the character of the buildings in the area; whereas on the fringe of the service area (i.e. at low arrival angles of the signal) the variation at 500 MHz has been found to be from 3 to 28 dB for different building densities.[309] At higher frequencies, channelling of radio energy along radial streets and improved reception at street intersections are common. The large variability of signal level from city to city in the USA is not entirely accounted for by differences in terrain.[256, 257] Differences in the median value within a small area (in the order of $0.25 \, \text{km}^2$) from the median for that distance show standard deviations to 3 to 12 dB.

In intermediate areas, where the rooftop level varies considerably, some attempt has been made to find a relationship between the degree of urbanisation and the transmission loss when the antenna heights are very much less than those of the buildings. At 450 MHz it has been found that deviations S(dB) from the median field-strength typical of urban areas can be correlated with the percentage $p_B\%$ of the area covered with buildings in the vicinity of the receiver as[258]

$$S = 30 - 20 \log p_B \quad \text{dB} \tag{7.16}$$

for $3\% < p_B < 50\%$. Values of S have been recorded within the range ± 20 dB. Such a relationship is necessarily an approximate guide since there will be very considerable variability according to

(*a*) the type of buildings present
(*b*) the orientation and precise location of individual buildings and
(*c*) the orientation of streets.

For the data on which eqn. 7.15 is based, all the measured values lay with ± 20 dB of the mean relationship.

Further corrections may be made to the mean prediction curves to allow for *height-gain;* that is, the change of measured field strength with increasing antenna height. For instance, the height-gain G_H (dB) realised by a rise from 3 to 10 m above ground is found to be 4 dB for frequencies of 40 to 100 MHz in hilly or flat terrain for both urban and rural areas, and 7 dB for about 200 MHz for flat terrain in rural areas (11 dB in urban or hilly areas). At 470 to 960 MHz it is 14 dB in urban areas, and is a function of the terrain roughness factor Δh(m) in rural areas, such that

$$G_H = 16 - 6 \log \Delta h \quad \text{dB} \tag{7.17}$$

but with a maximum value of 10 dB.[302] These factors apply for path lengths up to 50 km; for distances in excess of 100 km the values should be halved, with linear interpolation for intermediate distances. The height-gain figures quoted here are median values. At any specific location in an area, the actual height-gain may differ by many decibels from the median. The particular interest in G_H is that it facilitates a comparison between data collected for antennae 10 m above ground (a typical height for broadcast reception) and 3 m above ground (as may be used for vehicles). In areas with a high density of buildings, height-gain occurs in excess of the figures mentioned above, e.g. the median value is approximately 12 dB where the height of the antenna is doubled.[309] This figure applies below roof-top height. Above this level the height-gain is closer to 6 dB for doubling the height of the antenna above the general roof level,[259] as is to be expected from eqn. 4.16.

When using low antenna heights on an open (flat relfecting plane) site, the height-gain is not the same for vertical and horizontal polarisations. Over good (highly reflecting) ground, the effective antenna gain no longer decreases for vertical polarisation (see eqns. 4.17 to 4.19), whereas for horizontal polarisation there is a nearly linear height-gain even close to ground level. Over poor ground (with little reflectivity) the difference is less marked. In addition to this height-gain consideration, a vertical-polarisation antenna is much preferable to a horizontal-polarisation antenna for mobile communications, since the former may be a short vertical antenna which is omnidirectional and simple in both design and use.

In some broadcasting and land-mobile applications it is important to know the losses that occur with one terminal or both *within buildings*. Measurements made between 35 and 1500 MHz with distant high-sited transmitters have shown that field strengths at ground level (2m) inside buildings are about 15 to 25 dB less than that outside the building at ground level.[260-263] There are local variations about these values which are approximately log-normally distributed with a standard deviation of 8 to 14 dB. This 'building loss' decreases as the receiver is moved up inside the building, and at a height of approximately 30 m the received signal level has been found equal to that in surrounding streets 2 m above ground level. Measurements at 600 MHz on the ground floor and in the lofts of two-storey suburban houses have shown a mean attenuation of 19 dB and 10 dB, respectively, compared with measurements at 10 m in the street.[255]

The mechanism by which radiowaves propagate within large buildings is not fully understood. There is considerable variation in the basic construction and fittings of buildings, e.g. main frame, partitioning, ventillation shafts, wiring ducts, lifts and escallators. All of these affect the coupling of energy and re-radiation through the building. Consequently, no reasonable prediction technique can be attempted. Because of the large number of paths that contribute to the field at a point, the polarisation is likely to be random, and the standing wave pattern of peaks and nulls of field will be very complicated. Median levels inside buildings are essentially the same at 190 and

450 MHz, except that VHF is better at ground-floor level. Loss measurement within buildings showed 0·02 and 0·05 dB/m at 44 and 60 MHz, respectively.[264] Spatial fading within the building is more pronounced at UHF than VHF, i.e. positioning of the antenna is more critical. In principle, an antenna used for broadcast reception in a building may be positioned near a local field-strength maximum. However, because coupling radio energy into buildings is inefficient and the fading pattern inside is so marked, it may be that for establishing reliable communications with a man moving through a building (for emergency, security or other purposes), it will be necessary to set up a 'talk-through' intermediary station (such as a vehicle) at street level outside the building.

It is of interest that measurements for vehicles passing through long tunnels and under elevated roads showed much less transmission loss at 860 MHz than at 450 MHz.[265] This was attributed to more efficient launching of the wave into the tunnels and multipath reflections along the tunnel walls at the higher frequency.

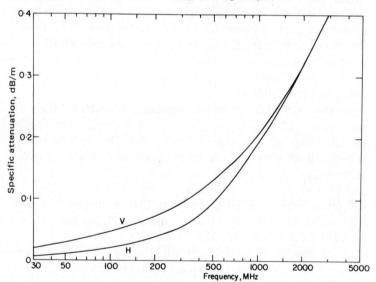

Fig. 7.9. *Attenuation through woodland*
V: Vertical polarisation
H: Horizontal polarisation

[Courtesy CCIR[307]]

Returning to the rural scene, the specific attenuation (dB/m) measured *through woodland* varies in the manner indicated in Fig. 7.9. These data apply for points in, or very close to, the woodland. In the

Figure, there is noticeably greater loss for vertical polarisation than for horizontal at frequencies less than about 1000 MHz. The attenuation may increase even more when the vegetation is particularly dense or the foliage wet. Measurements made at UHF behind deciduous woods in summer and winter have shown that the additional attenuation due to foliage is considerably less than that due to the bare trees alone.[255] In practice it has been observed that attenuation due to woodland does not exceed values of the order of 30 dB for frequencies below about 500 MHz.[309] At this upper frequency the specific attenuation is 0·1 dB/m, which implies that communication beyond 300 m is made possible by energy travelling along the top of the woodland and leaking downwards. On this assumption, the vegetation of dense forests and jungles over level ground has been modelled as a lossy dielectric slab.[266-268] The slab thickness is the effective height of the vegetation, and is characterised by a semi-empirical mean value of its complex permittivity ϵ'. For a propagation path of length d within the slab, and for a free-space wavelength λ, the transmission loss contains the term $\exp\left[-(2\pi/\lambda)d\sqrt{\epsilon'}\right]$, which is the specific attenuation given in Figure 7.9 based on earlier work.[269-272] However, the 'lateral wave' along the top of the slab is the only significant propagation mechanism for any but the shortest distances, even if both the transmitter and receiver are within the forest. The transmission loss is then proportional to d^4.

In the case where extensive woodland is situated between the transmitter and receiver, but both are well outside the woodland, the latter may act as a wide diffracting slab. On the other hand, for a thin belt of trees, the transmitted signal may be greater than the diffracted signal. For instance, for a bank of trees 1 km from a receiver and 10 m above the direct-ray line, the losses due to transmission and diffraction both equal about 10 dB at UHF if the bank of trees is 50 m thick. At VHF the diffraction and transmission losses are equal (at 8 dB) if the thickness is about 200 m.

A consequence of the field strength at a point in between buildings (or close to trees or other obstacles) being the sum of several contributions scattered off different surfaces, is that some *depolarisation* occurs. The cross-polarisation isolation (or polarisation discrimination, see Section 4.7) has been found to be a function of frequency, being about 18 dB at 35 MHz, but only 7 dB at 950 MHz when both antennae are at heights less than 10 m.[273] It is log-normally distributed, and the average value of the interdecile range (10% to 90%) in the frequency range 30 to 1000 MHz is about 15 dB, this being almost independent of whether the original polarisation was vertical or horizontal. However, changes of ground conductivity in wet weather are found to be important, particularly at low frequencies, and the motion of trees at quite moderate wind velocities also produces a decrease in cross-polarisation discrimination of several decibels.

For a transmitter antenna on a clear site, and receiving antenna at rooftop level in urban areas, the median cross-polarisation isolation may be 18 dB with levels of 9 and 29 dB for 90% and 10% of receiving sites.[309] The isolation is better in open country and worse at cluttered receiving sites for low antenna heights or where reception is poor. From a study of the type of terrain in the vicinity of the receiving antenna, data have been given on the way in which cross-polarisation isolation at 570 MHz decreases with increased diffraction loss in a suburban area, and how a wooded environment causes much greater depolarisation than suburban areas with a similar path loss.[309] The effect of radio frequency and building density was found to be negligible, whereas the effect of variation in the terrain roughness parameter Δh was very considerable.

7.3.3 Clutter factor method

In addition to the sophisticated prediction methods using a computer and a terrain data bank, and generally applicable prediction curves with their various correction factors, considerable attention has been given to relationships expressing the transmission loss L for a *smooth earth* (eqn. 4.16) in terms of the antenna heights h_t and h_r, the antenna gains G_t and G_r and the path length d, with a clutter factor β less than unity, such that

$$K = \frac{1}{L} = \frac{P_r}{P_t} = G_t G_r \left(\frac{h_t h_r}{d^2}\right)^2 \beta \qquad (7.18)$$

where K is the transmission factor. The factor β includes the losses (compared with the case of a smooth plane earth) due to terrain irregularity, buildings and trees causing shadowing, absorption and scattering of the radio energy. Although these losses vary widely with location, some success has been experienced with this approach.[131, 253] It has met with criticism because for most cases in which it is used there is no resemblance to the situation for which it strictly applies, i.e. line-of-sight propagation above a reflecting plane surface (for which $\beta = 1$). However, use of equation 7.18 is consistent with CCIR prediction methods mentioned in the last Section, since, for short ranges, those curves are approximately equivalent to $G_t G_r (h_t h_r / d^2)^2$; and the corrections for terrain morphology, vegetation and man-made structures

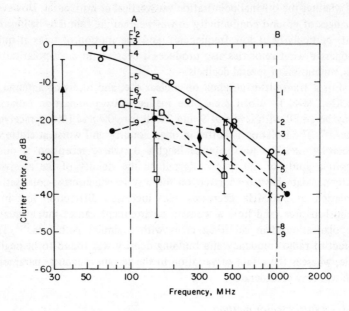

Fig. 7.10. *Variation of clutter factor β with frequency (for high transmitter and low receiver ≃ 3 m)*

Experimental studies:

o Rural[131]	◇ Rural[278]
x New York City[275]	◆ New York City[277]
● London[276]	⊓ 3 UK Cities[279]
⊓ New York City[277]	▲ Boston[259]

From CCIR prediction curves for 50% locations and 50% time[318]

△ Urban, $h = 1.5$ m

□ Rural, $h = 3$ m, $\Delta h = 50$ m

From CCIR corrections to overland prediction curves for standard deviation variation with location σ_L or time σ_T, terrain irregularity parameter Δh and path length d:

A: at VHF (plotted at 100 MHz)

1 $+ \sigma_T, d = 150$ km
2 $+ \sigma_L$
3 $\Delta h = 10$ m, $d < 100$ km
4 $+ \sigma_T, d = 50$ km
5 $- \sigma_T, d = 50$ km
6 $\Delta h = 150$ m, $d < 100$ km
7 $- \sigma_L$
8 $- \sigma_T, d = 150$ km
9 $\Delta h = 300$ m, $d < 100$ km

B: at UHF (plotted at 1000 MHz)

1 $+ \sigma_L, \Delta h = 300$ m
2 $\begin{cases} + \sigma_L, \Delta h = 50 \text{ m} \\ \Delta h = 10 \text{ m}, d < 100 \text{ km} \end{cases}$
3 $+ \sigma_T, d = 150$ km
4 $+ \sigma_T, d = 50$ km
5 $- \sigma_T, d = 50$ km
6 $- \sigma_T, d = 150$ km
7 $\begin{cases} \Delta h = 150 \text{ m}, d < 100 \text{ km} \\ - \sigma_L, \Delta h = 50 \text{ m} \end{cases}$
8 $- \sigma_L, \Delta h = 300$ m
9 $\Delta h = 300$ m, $d < 150$ km

may be regarded as equivalent to using a clutter factor. To be more precise, for 30 to 250 MHz and for 450 to 1000 MHz over land, the CCIR prediction curves show a distance dependence for ranges less than about 50 km closer to d^{-5} than the d^{-4} of eqn. 7.18; whereas for 450 to 1000 MHz on over-sea paths the relationship is closer to d^{-2}, as for free space. As was mentioned in the last Section, the h_r^2 relationship is found to be appropriate in open conditions or above roof-top level, but the height-gain is greater than this for antennae below roof-top level. Some authors have suggested that the urban clutter loss is a function of the angle of elevation of the energy incident on the receiving terminal.[252, 274] This may take account of the height-gain for terminals below roof-top level being greater than that above roof-top level.

Measured values of β, particularly those for city and suburban areas, have been reviewed by Longley,[253] and some of these for city areas (expressed in decibels) have been plotted in Fig. 7.10 as a function of frequency together with data for rural areas. Many other data have been reported, but they are not readily comparable.[253] The clutter factor for suburban areas is between city and rural values. The Figure also shows values of β calculated from the three available sets of CCIR prediction curves using low (mobile) terminals (but high base terminals).[318] Here β is determined by equating the transmission loss of eqn. 7.18 with the product of the transmission loss for free space (as given by eqn. 1.3) and the CCIR predicted attenuation with respect to free space. It is noteworthy that the CCIR data for rural and urban areas closely fit the measured data for rural areas, but that the city data show much more loss due to clutter. Compared with rural areas, the clutter factor for city areas seems to be about 20 dB at 100 MHz and 10 dB at 1000 MHz. This relative decrease at the higher frequency may be due to increased scatter off features of the taller buildings. To some extent the general increase with frequency of the clutter loss can be offset by utilising the increased antenna gain available at the higher frequencies.

Fig. 7.10 also includes variation from the median values for VHF and UHF as recommended by CCIR and taken from the text of Section 7.3.2. These are the increased clutter factors when the terrain irregularity parameter Δh is greater than 50 m, and the standard deviations σ_L and σ_T for different locations in hilly terrain and different times of sampling. A difficulty in applying these correction factors is that they are interdependent. Where a service must be provided in rural or urban areas with a range of terrain roughness and building density, it may not be useful to separate the various correction factors, but rather to prepare for the most unfavourable situation.

It is unlikely that the clutter factor would be used for predicting high signal levels (except perhaps for prediction of interference levels, see Section 8.3.2), but for small proportions of time and locations, and for high antennae, eqn. 7.18 may predict a transmission loss less than the free-space value. The free-space value should then be regarded as the upper limit.

7.4 Special problems with land-mobile terminals

A moving terminal (e.g. on a vehicle or carried by a man on foot) experiences a *slow variation* of signal strength which is clearly associated with changes in the type of area (e.g. hilltop or valley bottom, high building density or woodland etc.). In addition there is a very *rapid fading* due to the changing *relative phase* of signal components scattered or specularly reflected off various obstacles (e.g. buildings, trees and vehicles) and diffracted around them. The separation of signal peaks is comparable to the radio wavelength. In rural and even suburban areas this rapid (multipath-type) fading is less severe than in urban areas. As well as this very rapid fading and the slow change with type of area, there is an *intermediate fading* rate due to the localised changes of the *relative amplitude* of the different signal components as the aspect presented by buildings and other features changes. Both the very rapid and the intermediate rates of fading associated with position are indistinguishable from fading with time caused by the movement of nearby trees or vehicles.

For the slow and intermediate rate of fading that occurs with changes of position of a mobile terminal, the 'local median' is usually log-normally distributed with a standard deviation of 6 to 9 dB in the VHF and UHF bands.[131, 251] The statistical properties of the fast fading, corresponding to positional changes comparable with the wavelength, may be explained by assuming that there are many propagation paths between transmitting and receiving antennae, especially in a complex urban environment.[280] If all relative phases are assumed to be equally probable, the probability density functions of the amplitude and phase of the received signal may be evaluated as the mobile terminal moves along a street. Providing there is no direct ray between transmitter and receiver, the amplitude is Rayleigh distributed within an area for which the local median does not vary substantially. In particular, this condition is likely to be met at large distances from the transmitter. For a vehicle moving at about 60 km/h, the mean fading rate for a radio frequency of 90 MHz would be about 10 Hz, while for a frequency of 450 MHz this would rise to about 50 Hz (assuming two fades per free-space wavelength). It is important to note that the effects of slow and fast fading cannot be seprated by analysis of recordings of signal level, because the lower frequency components of the fast fading are indistinguishable from the slow fading.

The rapid and deep fading (up to 40 dB) associated with a vehicle moving through a spatially varying field pattern can be reduced by the use of frequency and/or space *diversity* techniques (as described in principle in earlier Chapters). Space diversity will require an antenna separation of more than a quarter wavelength. This is readily achieved at UHF and at the upper end of the VHF band. To achieve the same degree of decorrelation at the base station, the antenna separation has to be many wavelengths.[281] Numerous methods of implementing diversity techniques have been considered for mobile radio communication,[257, 282, 283] and 10 to 20 dB of transmitter power may be saved by these means.[284] As with the previously mentioned applications, the limitations of frequency bandwidth available for allocation make frequency diversity unattractive.

Another effect which may be significant for digital communications using mobile terminals is the limitation to *coherence bandwidth* due to the spread of multipath delays which occur. It can be shown that, for an envelope amplitude correlation of 0·5, the coherence bandwidth is $(2\pi\delta)^{-1}$, where δ is the time delay spread, while for a similar phase correlation the coherence bandwidth is $(4\pi\delta)^{-1}$.[257] It has been established experimentally in the UHF band that, for distant ranges from the transmitter (i.e. in areas where there is fast spatial fading), the excess delays follow an exponetial probability density function with delays in the region 0·25 to 5·0 μs.[285] These delays correspond to amplitude coherence bandwidths of 600 and 30 kHz, and phase coherence bandwidths of 300 and 15 kHz, respectively. Since the multipath delays are to a very large extent independent of frequency, the coherence bandwidths are applicable to both UHF and VHF operation. The resulting inter-symbol interference may be significant at high bit rates, although these would appear to present no problem at rates less than 20 kbit/s.

By using directional antennae on a vehicle, and so reducing the number of components to the received field, it may be possible to reduce considerably the amplitude of the fast fading and also the spread of path delay times, thereby increasing the coherence bandwidth. At 840 MHz, this has been found to be effective with antennae directed along or at right angles to the direction of motion (or in the direction of the transmitter.[286, 287] Orientation of the antenna had little effect on the mean attenuation level. In addition, the spread of the *Doppler spectrum* may be reduced, particularly with a sideways-looking antenna, since this spread is determined by the carrier frequency and the vehicle speed relative to the incoming wave. Signals arriving by various paths may be shifted in frequency by different amounts, and beats may occur between the resulting frequencies. In practice, the Doppler effect is not likely to be serious at VHF and UHF, although at 3·7 GHz it has been found to degrade intelligibility.[275]

The preceeding discussion has assumed transmission loss variation due to the movement of only one terminal. For the case of communication between two mobile terminals, the fading depths due to movement may be $\sqrt{2}$ times those when one terminal is moving, and the associated rate of fading and Doppler frequency may be greater or less than when one terminal is moving.

Chapter 8

Radio interference

8.1 Introduction

Although the prime object of radio engineering is to have a signal pass reliably from one place to another, this is of no use if the received signal is unintelligible due to interference or if the transmission causes interference with other systems.

The main interference to communication arises from sharing of the frequency spectrum between users in order to meet their requirements for bandwidth. Congestion of the frequency spectrum generally leads to potential interference between terrestrial systems, terrestrial and earth-to-space systems and between different earth-to-space systems. For instance, a frequency band close to 4 GHz is allocated by the ITU to both earth-to-satellite services and terrestrial fixed services, and specific frequencies within this band may be allocated to several users within a country by the proper authority in each country. Considerations of potential interference between users have led to detailed procedures to ensure that planned new radio links do not cause interference to existing links. 'Co-ordination distances' have been established between link terminals as a guide for isolating possible sources of interference.[325] This co-ordination distance is the minimum distance between transmitters and receivers of separate radio communication systems necessary to avoid interference for all but a given small time percentage. In order to determine the co-ordination distance for a specific source of interference, it may be necessary to examine a number of propagation mechanisms each of which may produce high signal levels (e.g. ducting, rain scatter, aircraft scatter etc.) in order to determine which of these is dominant for the radio paths involved.

Because of the need to share frequencies, some interference between systems may be inevitable, but the acceptable level of interference depends on the service in question, and perhaps on general propagation conditions too. For instance, lower levels of interference isolation (the ratio of the wanted to unwanted signal) are more acceptable for TV viewing during periods of low signal level than during period of normal signal level. In the former case, the system noise interference is relatively high, and so some reduction in co-channel isolation will not further degrade the quality of the picture.

A fundamental source of interference to communication is noise generated within the receiver and its antenna system. In addition, the receiving system may be subject to galactic noise, atmospheric noise due to lightning and thermal noise from clouds or rain, but man-made noise is usually more important than these. The topic of noise interference is considered in Section 8.2.

8.2 Noise interference

Determination of the minimum signal level required for satisfactory radio reception requires a knowledge of unwanted background noise levels. The term *'white noise'* is used because this consists of pulses which, by being randomly distributed in time and intensity, produce a continuous spectrum of frequencies when examined over a period of time. White noise energy emanating from different sources is summed in a receiver to give a total noise output. Some of this noise originates in the receiving system itself and some externally to the antenna. The latter may be of atmospheric, galactic (or cosmic) and man-made origins. An exception to white noise may occur when a receiver is close to certain electrical machinery, in which case specific frequencies may be present in the noise. Atmospheric noise, which is due to electrical discharges associated with thunderstorms, is unlikely to predominate at frequencies above about 30 MHz and is not considered further in this Section.

There is no single satisfactory index of noise interference for all types of radio service, but the *mean noise power* is the most generally useful. The noise power received from sources external to the receiver is conveniently expressed in terms of an *effective antenna noise factor* f_a, which is defined by

$$f_a = P_n/kT_0B = T_a/T_0 \qquad (8.1)$$

where P_n(W) is the mean noise power available from an equivalent loss-free antenna, k is Boltzmann's constant $(1·38 \times 10^{-23} \text{W/K/Hz})$, T_0 is a reference temperature (usually taken as about 290 K), B(Hz) is the effective receiver noise bandwidth over which energy is collected (approximately the 3 dB passband of the receiver) and T_a is the *effective antenna temperature* in the presence of external noise. Eqns. 8.1 illustrate two alternative methods of specifying the noise power, i.e. by the effective noise factor or by the effective temperature of the antenna. Both f_a and T_a are independent of bandwidth because the available noise power from all sources P_n may be assumed to be proportional to bandwidth.

It is difficult to comment usefully on noise levels in *receiving equipment* since developments are so rapid. Normally, the receiver noise factor is a few decibels, i.e. the noise temperature is well above ambient, but for communication systems which use a receiver with a noise temperature well below the ambient (e.g. a cooled receiver as used in an earth station may have a noise temperature of about 25 K), the noise power introduced by losses in the antenna feed line (as determined by eqn. 3.17) may be significant. Noise from the antenna itself is normally not a problem at VHF and above, since the ohmic resistance of an antenna is very much less than its radiation resistance. However, significant noise power may be introduced from the antenna side lobes, or the primary feed itself, pointing at a noise source such as the ground.

Of the sources of noise external to the antenna, *man-made noise* is dominant at VHF and higher frequencies for all but the quietest of rural areas. Such noise can arise from a number of sources, such as power lines, industrial electrical machinery, vehicle ignition systems etc., and it varies markedly with location and time.[288-290] Data are somewhat limited, but median values of man-made noise power, expressed in terms of F_a (dB above thermal noise at 290 K), are shown in Fig. 8.1. Curves *a* to *c*, for business, residential and rural areas were prepared from measurements made at 103 locations in the USA.[291, 292] 'Business areas' were defined as the core centres of large cities, 'residential areas' as the residential sections of large cities as well as the suburban areas of large population centres, and 'rural areas' as small communities and farms. It appears that the statistical variation from location to location about the median values of noise power has no clear dependence on frequency, but that the variation may be expressed as a standard deviation of 7 dB for the business area, 5 dB for the residential area and 6·5 dB for the rural area. Measurements made in the UK indicate that the man-made noise powers may be some 7–10 dB below those given in Fig. 8.1 at VHF.[310] Generally, it is found that the vertical component of the field due to man-made noise is stronger than the horizontal one.

Galactic noise which is not associated with particular radio-star sources or the sun is somewhat less than man-made noise, as may be seen from Fig. 8.1. However, above this background level, there are intense sources from the centre of the galaxy, from radio stars and

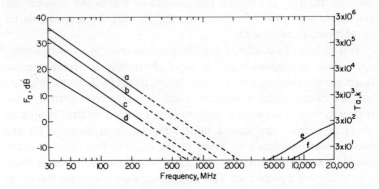

Fig. 8.1. *Man-made and galactic noise and that due to thermal emission in the atmosphere*

a Man-made noise in business area

b Man-made noise in residential area

c Man-made noise in rural area

d Galactic noise

e Typical noise due to rainfall and atmospheric absorption exceeded for 0·1% of the time in temperate latitudes at 30° elevation

f As *e*, but for 1·0% of the time

Effective antenna noise factor $F_a = 10 \log T_a/290$ (see text)

[Based on CCIR[305, 310]]

particularly from the sun, which has a noise temperature of about 10^6 K at 30 MHz and of at least 104 K at 10 GHz under quiet sun conditions. Large increases occur when the sun is disturbed (curves *a* to *d* of Fig. 8.1 apply for a short vertical loss-free grounded monopole antenna, and assume zero ground temperature).

In addition, noise power is generated from the *atmospheric gases, rain* and *clouds* (as has already been considered in general terms in Section 3.6), and this may restrict the use of certain communication channels. Curves given in Fig. 3.20 show the noise temperature as a function of frequency in the absence of cloud or rain. Above 10 GHz there is a progressive rise up to a maximum of about 290 K above 20 GHz due to the steadily increasing gaseous attenuation. Eqn. 3.15 relates the noise temperature with attenuation due to cloud or rain which may be important at frequencies above 5 GHz. For instance, a

3 dB attenuation due to rain (at 290 K) has an associated 140 K increase in noise temperature. Again, the maximum noise temperature value is the ambient temperature of the absorbing medium, which is shown as 290 K in Fig. 8.1.

The effect of noise external to the antenna depends to some extent on the *antenna directivity* (or gain). Clearly, if the noise power from outside the receiver and antenna system is essentially equally distributed with angle, the noise power received will be independent of the antenna gain, whereas the *wanted* signal power (and hence also the signal/noise ratio) will be proportional to the antenna gain. This will apply for galactic noise, but will not be generally true for man-made noise. In the case of fixed services from point-to-point, it may be possible to design the antennae to have nulls in the directions of any particular sources of man-made noise. To what extent clouds or rain cells fill the beam of an antenna will depend on the beamwidth, the distance from the antenna and the cloud or cell size. If the beamwidth is not filled, then the noise power (or effective noise temperature) will be less than if the beamwidth is filled.

8.3 Interference between co-channel services

8.3.1 General principles

In order to estimate mutual interference between two radio links it is necessary in principle to know the statistical distribution of the difference (in decibels) between the level of the interfering signal and that of the wanted signal. However, it may usually be assumed that the wanted signal does not experience any particular enhancement at the time when the unwanted signal is unusually high, and the problem may therefore be restricted to an estimation of signal level from a distant transmitter expected to be exceeded for only a small percentage of time. What this time percentage is will depend on the service under consideration.

The problem of co-channel interference is particularly acute in the case of a terrestrial fixed-point service (ground-to-ground) and an earth-space service at SHF, since the earth station transmits unusually high powers and receives unusually weak signals. Any coupling medium within the earth-space beam, notably rain, may scatter energy from the powerful transmission or to the sensitive receiver. Alternatively, atmospheric layering or ducting may cause high interference by way of the earth-station antenna side lobes. However, although interference between terrestrial and earth-space services may be the most sensitive problem, the content of this Section also applies to possible interference between two terrestrial links, be they both line-of-sight or troposcatter links, or one of each.

In the procedures recommended by the CCIR for assessing possible interference,[319, 325] scatter from rain or cloud is treated separately from all other ('clear air') propagation mechanisms likely to cause interference, because the former may lead to scatter well off the great-circle path, while the latter are effective only on the great-circle path. Rain or cloud scatter may be the dominant mechanism at SHF, especially for an earth station screened by local hills from interfering signals arriving by means of super refraction, though it is not of significance at VHF or UHF. Of the clear air propagation mechanisms, ducting is the most likely source of interference on transhorizon paths at SHF, though reflection from elevated layers in the atmosphere may be more important at VHF and the lower end of the UHF band.

Often one propagation mechanism may predominate to such a degree that the others may be ignored. This depends on the radio frequency, the nature of the path geometry, earth surface characteristics, climatic conditions or local terrain features at a path terminal. To assess whether interference could occur between one system and another, it is necessary to find what the predominant mechanism will be and whether the level will be significant.[319] However, because this procedure is somewhat lengthy, it is desirable to first use a quick procedure to determine whether an interference problem is likely,[325] and to then make a more detailed study only if this proves to be the case. Because of the risk of error in predicting transmission loss over short distances when the exact path geometry is unknown, detailed evaluation is always recommended if the potential source of interference is within 100 km. At frequencies above about 40 GHz, any large obstacle, such as a large building or a belt of thick trees or foliage, is effectively opaque. By careful location of terminals this feature can be used advantageously to reduce interference between systems.

In addition to clear air and rain scatter propagation mechanisms, energy reflected from the sporadic E layer of the ionosphere may also be a source of interference between services, but only at the lower end of the VHF band. This problem is discussed in Section 8.3.4.

8.3.2 Clear air effects
Determination of the potential interference levels between terminals which are on a *line-of-sight* path, with no obstacle obscuring the first Fresnel zone (see Section 4.2.1), is a relatively straightforward matter. It may be assumed that the basic transmission loss is equal to the free-

space loss added to any gaseous absorption losses, less the sum of instantaneous focussing effects and multipath interference between rays arriving over different paths in the atmosphere. The gaseous absorption losses may be estimated from the curves given in Fig. 3.14. The enhancement of signal due to focussing and multipath effects is found to be about 4·5 dB for 1% of time and 8·5 dB for 0·001% of time.[319] Even these small enhancement levels may be significant. They apply to paths of 50 km length and more, and for shorter distances they may be reduced proportionally.

The propagation mechanisms that may cause interference on paths *beyond line-of-sight* during clear air conditions, have a considerable dependence on frequency and on the time percentage of concern. For some applications interference may be tolerable for 10 or 20% of time, whereas for others it is not acceptable for more than 1 or even 0·1% of time. For the longer time percentages it is likely that the dominant mechanisms will be tropospheric scatter (see Section 2.7) whereas for the shorter time percentages partial reflections from atmospheric layers (see Section 2.6) may dominate at frequencies below about 1 GHz, and ducting may dominate above this frequency (see Section 2.5). In addition, very negative refractive index gradients reducing the diffraction loss on a path may be important for the longer or shorter time percentages, depending on the path distance and frequency (see Section 7.2 and especially Fig. 7.6).

For frequencies below about 1 GHz, the general methods outlined in Section 6.3 may be used to determine the enhancement $y_0(p)$ of the hourly-median signal level above the median value for time percentages p, where $y_0(p)$ is taken from Fig. 8.2 for signal enhancement above the median. As for eqn. 6.5, the factor $g(f)$ must be used to take account of frequency. Because the signal level exceeded for small time percentages is due to only a few contributions, possibly only one, the fading is very slow compared with that which applies for low-signal conditions when many contributions are present (see Section 6.4). For more than one contribution, with similar amplitudes and with the relative phases changing slowly, the fading has been described as 'roller-coaster fading'. The signal level remains almost constant except for deep nulls occurring during phase cancellation. Where detailed terrain data exists for the area surrounding a transmitter the potential levels of interference between that and one for which a site is being investigated may be determined by methods similar to those described in Section 7.3.1.[293]

In order to adapt general prediction methods to local conditions, and so predict local variability of transmission loss directly from meteorological data, attempts have been made to find some readily-obtainable *meteorological parameter* that is well correlated with transmission loss over transhorizon links. This is more easily achieved for predictions of high signal levels than for low signal levels. In some

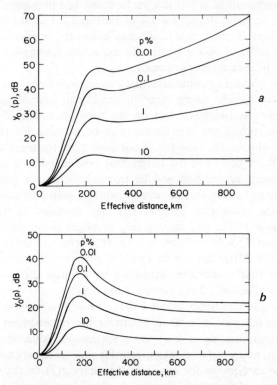

Fig. 8.2.   *Variation with effective distance of signal enhancement $y_0(p)$ above median exceeded for time percentages p*
a For an overland path in a maritime temperature climate (e.g. UK)
b For a continental climate (e.g. Continental Europe and USA)

[Courtesy CCIR[308]]

climates the transmission loss has been shown to have a high negative correlation with the surface refractive index N_s,[2, 294] and this has been found useful in predicting regional, seasonal and diurnal variations in transmission loss. In other climates this correlation has been found lacking.[295] Whether this correlation occurs may depend on whether there is a highly negative correlation between N_s and ΔN, the mean gradient of refractive index in the first kilometre above ground level. Some success in finding a relationship between transmission loss and refractive index gradients has been achieved by using the equivalent mean gradient between the ground and the common volume, and the gradient between the base of the common volume and 1 km above this base.[296, 308] Correlation with N_s is preferable to correlation with ΔN, or other gradients, since the former is readily obtained for any site while the latter requires the use of radiosondes. Unfortunately, weather stations where such radio sondes are used are widely spaced, typically by 250 km in Europe. This, together with the fact that seldom are more than two sondes launched per day, restricts the application of the method quite considerably.

There may be several reasons why ΔN may have a pronounced correlation with transmission loss. First, if ΔN falls well below its median value there will be an increase in atmospheric refraction, and a lowering of the height region at which scattering occurs to one where the refractive index fluctuations are greater. Secondly, the increased refraction will cause a reduction in the scattering angle, and the forward scattered power is an inverse function of the scatter angle. Thirdly, a highly negative ΔN may indicate

(a) the presence of one or more horizontally-stratified rapid decreases of refractive index with height somewhere within the first kilometre height, which may lead to ducting

(b) partial reflection or

(c) a large refractive index variance due to wind shear at the boundary.

Any of the above factors may lead to a considerably reduced transmission loss. It should be noted that where correlation of transmission loss with ΔN has been clearly demonstrated in temperate climates, it was possible only when the comparison was made on the basis of monthly, or at least weekly, medians.[297] Generally, there is no clear relationship between either N_s or ΔN and transmission loss in climates where the anuual range of these parameters is relatively small.

For frequencies above about 1 GHz, where *ducting* is expected to be the prime propagation mechanism for interference, predictions may be made on the basis of eqn. 2.35. Empirical values have been obtained for the parameters in this equation that may be expected for small time percentages.[319] However, the method is also valid for the case of strong super-refraction without ducting because of the similarity of the mathematical function involved in both cases. It is assumed that the transmission loss in excess of the line-of-sight attenuation can be expressed as a function of the path length, the minimum coupling distance into the duct, the rate of loss of energy from the duct with distance travelled, the climate, the radio frequency and the required percentage of time. The attenuation is increased if either or both terminals have a positive horizon elevation angle, and this increase may be equated to a single knife-edge diffraction loss with the assumption that the diffraction angle equals the sum of the horizon elevation angles.

Another cause of ('clear air') interference that occurs for small time percentages is reflection due to *aircraft*. Such interference is characterised by its short duration and high relative intensity, the coupling between the two antenna beams being effected by the aircraft itself. The probability of this occurring is dependent upon the frequency of aircraft movements through the intersections of the main beams or the minor lobes of the antennae.

8.3.3 Scatter by hydrometeors

Scatter from hydrometeors, particularly rain, may be a dominant cause of interference on transhorizon radio paths at frequencies above 1 GHz. It has the important characteristic that the scattering source may be *off the great-circle path*. Consequently, attention should be given to any directions for which the horizon angles as seen from the path terminals are low and for which the antenna polar diagrams would permit interference from hydrometeor scatter. Although as a first approximation rain scatter can be considered omnidirectional for vertical polarisation, there is an angular dependence for horizontally-polarised waves which is associated with the radiation pattern of the raindrops. Since these effectively act as dipole elements, the angular dependence has the form

$$P(\beta) = P_0 \cos^2 \beta \tag{8.2}$$

where β is the angle between the direction of scattering and the electric field in the incident wave. As a result, no power is scattered at right angles to the forward direction, and the scattered power at $45°$ off the forward (or backward) direction is half that scattered forwards (or backwards). Consequently, rain cells near to one terminal with a $90°$ scattering angle could produce very considerable interference for vertical polarisation but none for horizontal polarisation.

Although rain is the dominant source of hydrometeor scatter, there may often be a localised enhancement of scattering in the *melting layer* (or radar 'bright band' region, see Section 3.5). However, this layer may be expected to fill only a small proportion of the common volume of the two beams of the operational and interfering systems, compared with the volume filled by a raincell. Also, measurements have shown that scattering from *ice clouds* in the upper atmosphere may be a significant source of interference for small time percentages,[121] indeed, above about 15 GHz it may be more important than scattering from rain. The scattering cross-section of ice clouds increases with frequency at a greater rate than the associated attenuation (which is small), whereas the rate of increase of scattering due to rain becomes less than the attenuation. Furthermore, ice clouds are present for a longer proportion of the time than are rain cells.

When major portions of both main beams intersect in a rainstorm, interference may occur between stations that are well separated (i.e. by several hundred kilometres). When coupling occurs via main-lobe/side-lobe or side-lobe/side-lobe intersections, the interference is of consequence only at smaller distances, since the effective antenna gains are smaller. Furthermore, interference may occur at short distances (say less than 100 km) due to main-lobe/far-side-lobe intersection occurring in a heavy rainstorm.

General considerations have shown that, to a sufficient degree of accuracy, the *transmission loss L* for coupling due to scatter by rain is given by an equation of the form[298, 319]

$$L = \frac{K\lambda^2 r^2 A_g A_R BCP}{G_t ZMD} \tag{8.3}$$

where K is a constant, λ is the wavelength, r the distance from the terrestrial station to the scattering volume, A_g is an attenuation factor due to oxygen and water vapour gaseous absorption ($A_g = 1$ for frequencies less than about 3 GHz), A_R is a factor for attenuation within the rain, B is a factor for deviation from Rayleigh scattering ($B = 1$ for frequencies less than 10 GHz), C is a terrain blocking factor (e.g. half the beam of the terrestrial station may be obscured by the ground giving rise to 3 dB power loss), P is a factor allowing for the wave polarisaton and scattering angle β (such that $P = 1$ for a linear polarisation normal to the plane in which scattering occurs, and $P = \cos^2 \beta$ for a linear polarisation in this plane), G_t is the gain of the terrestrial station antenna, Z is the reflectivity factor for the height required (see Section 3.5), M is a polarisation mismatch factor to allow for the fact that the polarisation of the scattered interfering signal may be different from that of the receiving antenna (M is normally best taken as unity, i.e. the worst case so far as interference is concerned) and D is the effective dimension of the rain volume for scatter. The fact that the terms G_t and r for the terrestrial station are not matched by similar terms for the earth station (on the satellite path) is that rain is assumed to fill the narrow earth station beam for a length D, whereas the terrestrial beam is assumed relatively wide.

Table 8.1 Main ionospheric causes of interference at VHF

Cause of interference	Latitude zone	Period of severe interference	Approximate highest frequency with severe interference MHz	Approximate frequency above which interference is negligible MHz	Approximate range of distances affected kM E-W paths 3000–6000 or N-S paths 3000–10 000
Regular F-layer reflections	temperate	day, equinox-winter, solar-cycle maximum	50	60	
	low	afternoon to late evening, solar-cycle maximum	60	70	
Sporadic-E reflections	auroral	night	70	90	
	temperate	day and evening, summer	60	83–135*	500–4000
	equatorial	day	60	90	
Sporadic-E scatter	low	evening up to midnight	60	90	Up to 2000
Reflections from meteoric ionization	all	particularly during showers	may be important anywhere in the range		
Reflections from magnetic field aligned columns of auroral ionization	auroral	late afternoon and night			Up to 2000
Scattering in the F region	low	evening through midnight, equinox	60	80	1000–4000
Special transequatorial effects	low	evening through midnight	60	80	4000–9000

*

Scatter from rain is likely to be particularly important in monsoon climates, where the maximum height of raincells is almost double that for temperate climates and the rainfall rates are about three times those of temperate climates for any given time percentages less than 0·1%.

8.3.4 Ionospheric propagation

Although the ionosphere has little or no effect in producing co-channel interference above about 100 MHz, it can be of consequence in the VHF band below this frequency. 'sporadic E layer' propagation is the most significant ionospheric cause of interference, but long-distance transmission via the F2 layer can occur up to about 60 MHz for stations in temperate latitudes and 70 MHz for stations at low latitudes.[311] The most familiar form of the sporadic E layer occurs in an unpredictably random manner, but it is more likely to be present at certain times of day and months of the year than others according to location. It may be considered a negligible factor above 90 MHz except as a source of interference to circuits with high reliability requirements. Sporadic E propagation above 30 MHz rarely occurs at distances of less than 500 km, and the upper limit distance for one-hop propagation is about 2500 km. Multi-hop transmission at VHF via the sporadic E layer is possible, but it is unlikely. A detailed method has been developed for calculating the VHF field-strength due to sporadic E propagation for any percentage of time.[311] Studies made on 23 paths in Western Europe varying in length between 900 and 2500 km, and on five frequencies between 40 and 60 MHz, have shown maximum field-strengths occurring at distances of approximately 1400 to 1500 km.[299] Considerable variation was found from year to year, and the month of the maximum occurrences varied between May and August. There was no clear long-term correlation between solar activity and VHF propagation by Sporadic E. Table 8.1 shows the main causes of interference by ionospheric propagation at VHF.[311]

References

1 Radio regulations (Vols. I and II), ITU, Geneva, 1976
2 BEAN, B. R., and DUTTON, E. J.: *Radio Meteorology* (NBS Monograph No. 92 Supt. of Docs, US Govt. Printing Office, Washington DC, 1966)
3 HALL, M. P. M.: 'Variability of atmospheric water-vapour concentration and its implications for microwave-radio-link planning', *Electron Lett.*, 1977, **13**, pp. 650–651
4 UK Meteorological Office: 'World distribution of atmospheric water vapour pressure'. *Geophysical Memoirs, No. 100,* 1958, HMSO, London
5 HALL, M. P. M.: 'Radiosondes for radiometeorological research'. NATO Advanced Study Institute Proceedings. *'Statistical methods and instrumentation in geophysics',* 1971, Teknologisk Forlag, Oslo, Norway
6 CRAIN, C. M.: 'Apparatus for recording fluctuations in the refractive index of the atmosphere at 3.2 cm wavelength', *Rev. Sci. Ins.,* 1950, **21**, p. 456
7 BIRNBAUM, G.: 'A recording microwave refractometer', *ibid.,* 1950, **21**, p 169
8 VETTER, M. J., and THOMPSON, M. C., Jun.: 'An absolute microwave refractometer', *ibid.,* 1962, **33**, pp. 656–660
9 VETTER, M. J., and THOMPSON, M. C., Jun.: 'Direct reading microwave refractometer with quartz-crystal reference', *IEEE Trans.,* 1971, **IM-20**, p. 58
10 HAY, D. R., MARTIN, H. C., and TURNER, H. E.: 'A lightweight refractometer', *Rev. Sci. Ins.,* 1961, **32**, pp. 693–697
11 DEAM, A. P.: 'Radiosonde for atmospheric refractive index measurements', *ibid.,* 1962, **33**, p. 438
12 HALL, M. P. M., and COMER, C. M.: 'Statistics of tropospheric radio-refractive-index soundings taken over a 3-year period in the UK', *Proc. IEE,* 1969, **116**, pp. 685–690
13 BEAN, B. R., CAHOON, B. A., SAMSON, C. A., and THAYER, G. D.: *A world atlas of atmospheric radio refractivity,* (ESSA Monograph No. 1, [Access No. AD648–805], NTIS, Springfield, Va, USA, 1966)
14 BEAN, B. R., and THAYER, G. D.: 'On models of the atmospheric radio refractive index', *Proc. IRE,* 1959, **47**, pp. 740–755
15 CASTEL, F. (du) and MISME, P.: 'Eléments de radioclimatologie (Elements of radioclimatology)', *L'Onde Electrique,* 1957, **37**, pp. 1049–1052
16 LANE, J. A.: 'The radio refractive index gradient over the British Isles', *J. Atmos. & Terr. Phys.,* 1961, **21**, pp. 157–166

17 AKAYAMA, T.: 'Tropospheric radio-refractivity gradient over Japan and its vicinity', *J. Inst. Electron & Commun. Eng.*, 1971, **54-B**, p. 175

18 BEAN, B. R., HORN, J. D. and OZANICH, A. M., Jun.: *Climatic charts and data of the radio refractive index for the United States and the World* (NBS Monograph No. 22, Supt. of Docs, United States Government Printing Office, Washington, D.C., 1960)

19 BOOKER, H. G. and WALKINSHAW, W.: 'The mode theory of tropospheric refraction and its relation to waveguides and diffraction', *Meteorological factors in radio wave propagation*, Phys. Soc., 1947, pp. 80–127

20 HALL, M. P. M., and COMER, C. M.: 'Changes in radio field strength at VHF and UHF due to disintegration of reflecting layers in the troposphere', *Proc. IEE*, 1970, **117**, pp. 1925–1932

21 DOUGHERTY, H. T., and HART, B. A.: 'Anomalous propagation and interference fields' OT Report 76–107 (Access No. PB 262477), NTIS, Springfield, Va, USA, December 1976

22 CHANG, H. T.: 'The effect of tropospheric layer structures on long-range VHF radio propagation', *IEEE Trans.*, 1971, **AP-19**, pp. 751–756

23 WAIT, J. R.: *Electromagnetic waves in stratified media* (Pergamon Press, Oxford, 1962)

24 DOUGHERTY, H. T.: 'Recent progress in duct propagation predictions', *IEEE Trans.*, 1979, **AP-27**, pp. 542–548

25 WAIT, J. R., and SPIES, K. P.: 'Internal guiding of microwaves by an elevated tropospheric layer', *Radio Sci.*, 1969, **4**, pp. 319–326

26 CHO, S. H., and WAIT, J. R.: 'E M wave propagation in a laterally non-uniform troposphere', *ibid.*, 1978, **13**, p. 253

27 DOUGHERTY, H. T., McGAVIN, R. E., and KRINKI, R. W.: 'An experimental study of the atmospheric conditions conducive to high radio-fields'. OT Report (Access No. COM-75-11138/AS), NTIS, Springfield, Va, USA, September 1970

28 LANE, J. A.: 'Small-scale variations of radio refractive index in the troposphere. Pt. 1. Relationship to Meteorological conditions', *Proc. IEE*, 1968, **115**, pp. 1227–1234

29 GRAIN, C. M.: 'Survey of airborne refractometer measurements', *Proc. IRE*, 1955, **43**, p. 1405

30 BUSSEY, H. E., and BIRNBAUM, G.: 'Measurement of variation in atmospheric refractive index with an airborne microwave refractometer', *N.B.S. J. Res.*, 1953, **51**, pp. 171–178

31 CASTEL, F. (du): *Propagation tropospherique et faisceaux hertziens (Tropospheric propagation and transhorizon beams)* (Editions Chiron, Paris, 1961)

32 FRIIS, H. T., CRAWFORD, A. B., and HOGG, D. C.: 'A reflection theory for propagation beyond the horizon', *B.S.T.J.*, 1957, **36**, pp. 627–644

33 HALL, M. P. M.: 'Further evidence of VHF propagation by successive reflections from an elevated layer in the troposphere', *Proc. IEE*, 1968, **115**, pp. 1595–1596

34 WAIT, J. R.: 'Whispering-gallery modes in a tropospheric layer', *Electron Lett.*, 1968, **4**, pp. 377–378

35 CHO, S. H., and WAIT, J. R.: 'Analysis of microwave ducting in an inhomogeneous troposphere', *Pure & Appl. Geophys*, 1978, **116**, pp. 1118–1142

36 TATARSKI, V. I.: *Wave propagation in a turbulent medium*, (translated by R. A. SILVERMAN, McGraw-Hill, New York, 1961)

37 OTTERSTEN, H.: 'Atmospheric structure and radar backscattering in clear air', *Radio Sci.*, 1969, **4**, pp. 1179–1193

38 MATHUR, N. C. and RAMARO, C. T.: 'Rake probing of atmospheric refractive structure', *J. Inst. Electron. & Telecommun. Eng.*, March 1977

39 LANE, J. A., and PALTRIDGE, G. W.: 'Small-scale variatons of radio refractive index in the troposphere. Part II. Spectral characteristics', *Proc. IEE*, 1968, **115**, pp. 1235–1239

40 SAXTON, J. A., LANE, J. A., MEADOWS, R. W., and MATTHEWS, P. A.: 'Layer structure of the troposphere. Simultaneous radar and microwave refractometer investigations', *ibid.*, 1964, **111**, pp. 275–283

41 WATERMAN, A. T.: 'A rapid beam-swinging experiment in transhorizon propagation', *IRE Trans*, 1958, **AP-6**, pp. 338–340

42 MEADOWS, R. W.: 'Tropospheric scatter observations at 3480 Mc/s with aerials of variable spacing', *Proc. IEE*, 1961, **108B**, pp. 349–360

43 ZAVODY, A. M.: 'Effect of scattering by rain on radiometer measurements at millimetre wavelengths', *ibid.*, 1974, **121**, pp. 257–265

44 ROGERS, R. R.: 'The mesoscale structure of precipitation and space diversity', *J. Tech. Atmos*, 1974, **8**, pp. 485–490

45 HARROLD, T. W., and AUSTIN, P. M.: 'The structure of precipitation systems – a review', *J. Rech. Atmos.*, 1974, **VIII**, pp. 41–57

46 KREITZBERG, C. W., and BROWN, H. A.: 'Mesoscale weather systems with an occlusion', *J. Appl. Meteorol.*, 1970, **9**, pp. 417–432

47 ELLIOTT, R. D., and HOVIND, E. L.: 'On convective bands within pacific coast storms and their relation to storm structure', *ibid.*, 1964, **3**, pp. 143–154

48 HARROLD, T. W.: 'Mechanisms influencing the distribution of precipitation within baroclinic disturbances', *Q. J. R. Meteorol. Soc.*, 1973, **99**, pp. 232–251

49 OMOTO, Y.: 'On pre-frontal precipitation zones in the United States'. *J. Meteorol. Soc.* 1965, **43**, pp. 310–330

50 BROWNING, K. A. and HARROLD, T. W.: 'Air motion and precipitation growth in a wave depression'. *Q. J. R. Meteorol. Soc.*, 1969, **95**, pp. 288–309

51 HALL, M. P. M., and GODDARD, J. W. F.: 'Variation with height of radar reflectivity due to rain and the influence of wind shear on raincells', *Ann. Telecommun.*, 1977, **32**, pp. 444–448

52 CHERRY, S. M., GODDARD, J. W. F., HALL, M. P. M., and KENNEDY, G. R.: 'Measurement of raindrop-size distributions using dual-polarization radar', Proceedings of Fifth International Conference on Erosion by Solid and Liquid Impact, Cambridge, UK, September 1979

53 LIN, S. H.: 'A method for calculating rain attenuation distributions on microwave paths', *Bell Syst. Tech. J.*, 1975, **54**, pp. 1051–1086

54 MISME, P., and FIMBEL, J.: 'Détermination théorique et experimentale de l'affaiblissement par la pluie sur un trajet radioeléctrique (Theoretical and experimental determination of attenuation by rain on a radio path)' *Ann. Telecommun.*, 1975, **30**

55 RUE, O.: 'Effects of rainstorms on the system performance of multi-hop radio-relay links'. Proceedings of the URSI Symposium on Propagation in Non-Ionised Media, La Baule, France, 1977, pp. 311–316

56 MORITA, K., and HIGUTI, I.: 'Prediction methods for rain attenuation distributions of micro and millimetre waves', *Rev. Electr. Commun. Lab.*, 1976, **24**

57 BERTOK, E., DE RENZIS, G., and DRUFUCA, G.: 'Estimate of attenuation due to rain at 11 GHz from raingauge data' Proceedings of the URSI Symposium on propagation in non-ionised media, La Baule, France, 1977, pp. 295–300

58 DRUFUCA, G.: 'Rain attenuation statistics for frequencies above 10 GHz from rain-gauge observations', *J. Rechs. Atmos.*, 1974, **8**, pp. 399–411

59 HARDEN, B. H., NORBURY, J. R., and WHITE, W. J. K.: 'Measurements of rainfall for studies of millimetre radio attenuation', *IEE J. Microwave Opt. & Acoust.*, 1977, **1**, pp. 197–202

60 EVANS, H. W.: 'Attenuation on Earth-space paths at frequencies up to 30 GHz' ICC-71 Conference, Montreal, Canada, 1971

61 BRADLEY, J. H. S.: 'Rainfall extreme value statistics applied to microwave attenuation climatology'. McGill University Stormy Weather Group Scientific Report, MW-66, 1970

62 KINASE, A. *et al..:* 'Statistics of attenuation due to precipitation of radio waves in 12 GHz band at higher angles of elevation'. NHK Laboratories Note, Series 171, Japan, October 1973

63 DEBRUNNER, W. E., and LINIGER, M.: 'Einfluss der struktur von regengebieten auf, die ausbreitung von mikrowellen (Influences of the structure of rain areas on microwave propagation)', *PTT Techn. Mitteilungen*, 1975, **53**

64 SKERJANEC, R. E., and SAMSON, C. A.: 'Rain attenuation study for 15 GHz relay design'. Federal Aviation Admin. Report. FAA-RD-70-21, Washington. DC, 1970

65 STRICKLAND, J.I.: 'Radar measurements of site-diversity improvement during precipitation', *J. Rech. Atmos.*, 1977, **8**, p. 451

66 MARSHALL, J. S., and PALMER, W. M. K.: 'The distribution of raindrops with size', *J. Meteorol.*, 1948, **5**, pp. 165–166

67 ROGERS, R. R., and PILIE, R. J.: 'Radar measurements of drop-size distribution', *J. Atmos. Sci.*, 1962, **19**, pp. 503–506

68 PRUPACHER, H. R., and PITTER, R. L.: 'A semi-empirical determination of the shape of cloud and rain drops', *ibid.*, 1971, **28**, pp. 86–94

69 PRUPACHER, H. R., and BEARD, K. V.: 'A wind tunnel investigation of the internal circulation and shape of water drops falling at terminal velocity in air', *Q. J. R. Meteorol. Soc.*, 1970, **96**, pp. 247–256

70 SAUNDERS, M. J.: 'Cross-polarisation at 18 and 30 GHz due to rain', *IEEE Trans.*, 1971, **AP-19**, pp. 273–277

71 BRUSSAARD, G.: 'A meteorological model for rain-induced cross-polarisation', *ibid.*, 1976, **AP-24**, pp. 5–11

72 MARKOWITZ, A. H.: 'Raindrop size distribution expressions', *J. Appl. Meteorol.* 1976, **15**, pp. 1029–1031

73 ATLAS, D., and ULBRICH, C.W.: 'The physical basis for attenuation – rainfall relationships and the measurement of rainfall parameters by combined attenuation and radar methods', *J Rech. Atmos.*, 1974, **8**, pp. 275–298

74 GUNN, R., and KINZER, G. D.: 'The terminal velocity of fall for water droplets in stagnant air', *J. Meteorol.*, 1949, **6**, pp. 243–248

75 BEST, A. C.: 'Empirical formulae for the terminal velocity of water drops falling through the atmosphere', *Q. J. R. Meteorol. Soc.*, 1950, **76**, pp. 302–311

76 BEARD, K. V., and PRUPPACHER, H.R.: 'A determination of the terminal velocity and drag of small water drops by means of a wind tunnel', *J. Atmos. Sci.*, 1969, **26**, 1066–1072

77 BRADLEY, S. G., and SNOW, C. D.: 'The measurement of charge and size of raindrops: Part II, Results and analysis at ground level', *J. Appl. Meteorol.*, 1974, **13**, pp. 131–147

78 UK Meteorological Office, *Meteorological glossary* (Compiled by D. H. McINTOSH, HMSO, London, 1963)

79 RYDE, J. W.: 'Attenuation and radar echoes produced at centimetre wavelengths', *Meteorological factors in radio-wave propagation*, Phys. Soc., London, 1946

80 UK Meteorological Office: *Handbook of meteorological instruments. Pt I, Instruments for surface observations* (HMSO, London, 1961)

81 NORBURY, J. R., and WHITE, W. J. K.: 'A rapid-response raingauge', *J. Phys.* 1971, **4**, pp. 601–602

82 SEMPLAK, R. A.: A gauge for continuously measuring rate of rainfall, *Rev. Sci. Ins.*, 1966, **37**, pp. 1554–1558

83 SELIGA, T. A., and BRINGI, V. N.: 'Potential use of radar differential reflectivity measurements at orthogonal polarizations for measuring precipitation', *J. Appl. Meteorol.*, 1976, **15**, pp. 69–76

84 ECCLES, P. J.: 'Attenuation from dual-wavelength radar observations of hailstorms', Proceedings of the URSI Symposium on propagation in nonionized media, La Baule, France, 1977, pp. 511–514

85 HARDEN, B. N., NORBURY, J. R., and WHITE, W. J. K.: 'Estimation of attenuation by rain on terrestrial radio links in the UK at frequencies from 10 to 100 GHz', *IEE J. Microwave Opt. & Acoust.*, 1978, **2**, pp. 97–104

86 JOSS, J., and WALDVOGEL, A.: 'Raindrop size distribution and scattering size errors', *J. Atmos. Sci.*, 1969, **26**, pp. 566–569

87 DONNADIAN, G.: 'Etude des caractéristiques physiques et radioélectriques de la pluie a l'aide l'un spectropluviomètre photoélectrique (Studies of the physical and radioelectric characteristics of rain with the aid of a photoelectric raindrop spectrometer)', *J. Rech. Atmos.*, 1974, **VIII**, pp. 253–266

88 KNOLLENBERG, R. G.: 'The optical array: an alternative to scattering or extinction for airborne particle size determination', *J. Appl. Meteorol.*, 1970, **9**, pp. 86–103

89 RYDE, J. W., and RYDE, D.: 'Attenuation of centimetre waves by rain, hail, fog and clouds'. Research Report of General Electricity Co. Wembley, England, 1945

90 MEDHURST, R. C.: 'Rainfall attenuation of centimetre waves: Comparison of theory and measurement', *IEEE Trans.*, 1965, **AP-13**, pp. 550–564

91 SETZER, D. E.: 'Computed transmission through rain at microwave and visible frequencies', *Bell. Syst. Tech. J.*, 1970, **49**, pp. 1873–1892

92 GUNN, K. L. S., and EAST, T. W. R.: 'The microwave properties of precipitation particles', *Q. J. R. Meteorol., Soc.*, London, 1954, **80**, pp. 522–545

93 OLSEN, R. L., ROGERS, D. V., and HODGE, D. B.: 'The aR^b relation in the calculation of rain attenuation', *Trans. IEEE*, 1978, **AP-26**

94 LAWS, J. O., and PARSONS, D. A.: 'The relation of raindrop size to intensity', *Trans. Amer. Geophys. Union*, 1943, **24**, pp. 452–460

95 RAY, P. S.: 'Broadband complex refractive indices of ice and water', *Appl. Optics.*, 1972, **II**, pp. 1836–1844

96 CRANE, R. K.: 'Attenuation due to rain – A mini review', *IEEE Trans.*, 1975, **AP-23**, pp. 750–752

97 WALDTEUFEL, P.: 'Attenuation des ondes hyperfréquences par la pluie: une mise au point' (Attenuation of hyperfrequency waves by rain: as determined at a point)', *Ann. Telecommun.*, 1973, **28**, pp. 255–272

98 FEDI, F., and MANDARINI, P.: 'Analysis of the influence of the various parameters on the attenuation-rain rate relations. In *Modern topics in microwave propagation and air-sea interaction* (D. Reidel Publishing Co. Dordrecht, Holland, 1973)

99 SANDER, I.: 'Rain attenuation of millimetre waves at $\lambda = 5.77$, 3.3 and 2 mm, *IEEE Trans.*, 1975, **AP-23**, pp. 213–220

100 NORBURY, J. R., and WHITE, W. J. K.: 'Microwave attenuation at 35.8 GHz due to rainfall; *Electron. Lett.*, 1972, **8**, 91

101 CRANE, R. K.: 'The rain range experiment – propagation through a simulated rain environment', *IEEE Trans.*, 1974, **AP-22**, pp. 321–328

102 FEDI, *et al.*: 'Attenuation: theory and measurements', *J. Rech. Atmos.*, 1974, **8**, pp. 465–472

103 MORITA, K., *et al.*: 'Radio propagation characteristics due to rain at 20 GHz band', *Rev. Electr. Commun. Lab.*, 1974, **22**, p. 627

104 CHU, T. S.: 'Rain-induced cross polarization at centimeter and millimeter wavelengths', *Bell Syst. Tech. J.*, 1974, **53**, pp. 1568–1569

105 OGUCHI, T., and HOSOYA, Y.: 'Scattering properties of oblate raindrops and cross polarization of radio waves due to rain, (Part II) Calculations at microwave and millimeter wave regions', *J. Radio. Res. Labs. Japan*, 1974, **21**, pp. 191–259

106 OGUCHI, T.: 'Scattering properties of Pruppacher-and-Pitter form raindrops and cross-polarisation due to rain. Calculations at 11, 13, 19.3, 34.8 GHz', *Radio Sci.*, 1977, **12**, pp. 41–51

107 FIMBEL, J., JUY, M., and BOITHIAS, L.: 'Importants affaiblissements différentials dus à la pluie mesurés sur un liaison de 53 km à 13 GHz (Important attenuation due to rain measured on a path of 53 km at 13 GHz)', *Electron. Lett.*, 1976, **12**, pp. 119–120

108 FEDI, F., MERLO, U., and MIGLIORINI, P.: 'Effect of rain structure on rain-induced attenuation', *Ann. Telecommun.*, 1977, **32**, pp. 459–464

109 JOSS, J., THAMS, J. C., and WALDVOGEL, A.: 'The variation of raindrop size distributions at Locarno'. Proceedings of the International Conference on Cloud Plysics, Toronto, 1968, pp. 369–373

110 ECCLES, P. J., and MEULLER, E. A.: 'X-band attenuation and liquid water content estimation by a dual-wavelength radar', *J. Appl. Meteorol.*, 1971, **10**, pp. 1252–1259

111 CRANE, R. K.: 'Prediction of the effects of rain on satellite communication systems', *Proc. IEEE*, 1977, **65**, pp. 456–474

112 MISME, P.: 'Etude expérimentale de la propagation des ondes millimétriques dans les bandes de 5 et de 3 mm (Experimental study of EHF propagation in the 5 and 3 mm bands)', *Ann. Telecommun.*, 1966

113 AHMED, I. Y., and AUCHTERLONIE, L. J.: 'Microwave measurements on dust, using an open resonator', *Electron. Lett.*, 1976, **12**, p. 445

114 DEVASIRVATHAM, D. M. J., and HODGE, D. B.: 'Power law relationships for rain attenuation and reflectivity'. Ohio State University E.S.C. Technical report 784650-2, Department of Electrical Engineering, O.S.U., Columbus, Ohio. 43212, January 1978

115 BATTAN, L. J.: *Radar observation of the atmosphere* (University of Chicago Press, 1973)

116 JOSS, J., SCHRAM, K., THAMS, J. C., and WALDVOGEL, A.: *On the quantitative determination of precipitation by radar* (Wissenschaftliche Mitteilung Nr 63, Zurich, Eidgenossische kommission zum stadium der Hagelbildung und der Hagelabwehr)

117 STOUT, G. E., and MUELLER, E. A.: 'Survey of relationships between rainfall rate and radar reflectivity in the measurement of precipitation', *J. Appl. Meteorol.*, 1968, 7, pp. 465–474

118 YAMADA, M., OGAWA, A., FURUTA, O., and YOKOI, H.: 'Measurement of rain attenuation by dual-frequency radar'. International Symposium on Antennas and Propagation, Sendai, Japan, August 1978

119 HALL, M. P. M., and GODDARD, J. W. F. 'Variation with height of the statistics of radar reflectivity due to hydrometeors', *Electron Lett.*, 1978, 14, pp. 224–225

120 KATZ, I.: 'A rain cell model'. 17th Conference on radar meteorology, Am. Met. Soc. October 1976, pp. 442–447

112 OLSEN, R. L., and LAMMERS, U. H. W.: 'Bistatic radar measurements of ice-cloud reflectivities in the upper troposphere', *Electron Lett.*, 1978, 14, pp. 219–221

122 CRAWFORD, A. B., and JAKES, W. C.: 'Selective fading of microwaves', *Bell Syst. Tech. J.*, 1952, pp. 68–90

123 BOITHIAS, L., and BATTESTI, J.: 'Protection contre les évanouissements sur les faisceaux hertziens en visibilité (Protection against fading on line-of-sight radio-relay systems)', *Ann. Telecommun.*, 1967, 22, pp. 230–242

124 PICQUENARD, A.: *Radio wave propagation* (Macmillan Press Ltd., London, 1974)

125 BECKMANN, P., and SPIZZICHINO, A.: *The scattering of electromagnetic waves from rough surfaces* (Pergamon Press Ltd., Oxford, 1963)

126 KERR, D. E.: *Propagation of short radio waves* (McGraw-Hill Book Co. Inc., Radiation Laboratory Series, 1951, Vol. 13, pp. 9–22)

127 NORTON, K. A., and OMBERG, A. C.: 'Maximum range of a radar set', *Proc. IRE*, 1947, 35, p. 4

128 BURROWS, C. R., and ATWOOD, S. S.: *Radiowave propagation* (Academic Press, 1949)

129 BULLINGTON, K.: 'Radio propagation fundamentals', *Bell Syst. Tech. J.*, 1957, 36, pp. 593–626

130 WOJNAR, A.: 'Ground-wave radio links: unified analysis in simple terms', Eurocon '77, 1977, Vol. 1, paper 2, 3.6, p. 262

131 EGLI, J.: 'Radio propagation over 40 Mc/s over irregular terrain', *Proc. IRE*, 1957, 45, pp. 1382–1391

132 NORTON, K. A., HUFFORD, G. A., DOUGHERTY, H. T., and WILKERSON, R. E.: 'Diversity design for within-the-horizon radio relay systems' NBS Report 8787 (Access No. COM-73-10103), NTIS, Springfield, Va, USA, 1965

133 BOITHIAS, L.: *Calcul par nomogrammes de la propagation des ondes* (Four Language Edition, Eyrolles, Paris, 1972)

134 BRAMLEY, E. N., and CHERRY, S. M.: 'Investigation of microwave scattering by tall buildings', *Proc. IEE*, 1973, **120**, pp. 833–842

135 CARTWRIGHT, N. E., and TATTERSALL, R. L. O.: 'Simultaneous measurements of radio refractivity and multipath fading on 2nd July 1975, at 11, 19 and 36 GHz on a 7.5 km path', *Electron. Lett.*, 1977, **13**, pp. 208–210

136 SMITH-ROSE, R. L., and STICKLAND, A. C.: 'An experimental study of the effect of meteorological conditions upon the propagation of centimetre radio waves', *Meteorological factors in radio wave propagation.*, Phys. Soc. London, 1946

137 VIGANTS, A.: 'Space-diversity engineering', *Bell, Syst. Tech. J.*, 1975, **54**

138 FEHLHABER, L.: 'Influence of the path geometry on fading on line-of-sight radio-relay paths', *Tech. Ber. Des Forschungs Instituts Der DBP beim FTZ*, 1976, **455**, p. 59

139 DRUFUCA, G., and TORLASCHI, E.: 'Rain outage performance of tandem and route diversity systems at 11 GHz', *Rad. Sci.*, 1977, **12**, pp. 63–74

140 VALENTINE, R.: 'Attenuation caused by rain at frequencies above 10 GHz', *Ann. Telecommun.*, 1977, **32**, pp. 465–468

141 HARDEN, B. N., NORBURY, J. R., and WHITE, W. J. K.: 'Measurements of point and line rainfall rates for microwave attenuation studies' Proceedings of the URSI Symposium on propagation in non-ionized media, La Baule, France, 1977, pp. 279–281

142 DAMOSSO, E. D., and PADOVA, S de.: 'Rain attenuation study at 11 GHz. Experimental results and investigation on the statistical properties of single attenuative events', *Ann. Telecommun.*, 1977, **32**, pp. 449–453

143 BATTESTI, J., BOITHIAS, L., and MISME, P.: 'Determination of attenuation due to rain for frequencies above 10 GHz', *ibid.*, 1971

144 MORITA, K., and HIGUTI, I.: 'Statistical studies on electromagnetic wave attenuation due to rain', *Rev. Electr. Commun. Lab.*, 1971, **7–8**, p. 798

145 BUSSEY, H. E.: 'Microwave attenuation statistics estimated from rainfall and water vapor statistics', *Proc. IRE*, 1950, **38**, pp. 781–785

146 MORITA, K.: 'Prediction of Rayleigh fading occurrence probability of line-of-sight microwave links'. *Rev. Electr. Commun. Lab.*, 1970, **18**, pp. 11–12

147 TATARSKI, V. I.: *The effects of the turbulent atmosphere on wave propagation* (Nauka, Moscow, 1967)

148 BULLINGTON, K.: 'Phase and amplitude variations in multipath fading of microwave signals', *Bell. Syst. Tech. J.*, 1971, **50**

149 LIN, S. H.: 'Statistical behaviour of a fading singal', *ibid.*, 1971, **50**, pp. 3211–3270

150 MOGENSEN, G.: 'Experimental investigations of radio wave propagation in the 13.5 to 15.0 GHz frequency band'. Publication Number LD 30, Electromagnetics Institute, Technical University of Denmark, April, 1977

151 ROORYCK, M., and JUY, M.: 'Résultats de 5 ans de mesures à 13 GHz sur un trajet de 53 km (Results of 5 years of measurements at 13 GHz on a 53 km path)' Note Technique TCR-APH-43, CNET, Paris, 1977

152 ALLNUTT, J. E.: 'Nature of space diversity in microwave communications via geostationary satellites: a review', *Proc. IEE*, 1978, **125**, pp. 369–376

153 MARINO, H., and MORITA, K.: 'Design of space diversity receiving and transmitting systems for line-of-sight microwave links, *IEEE Trans.*, 1967, **COM-15**, pp. 603–614

154 MARTIN-ROYLE, R. D., and DUDLEY, L. W.: 'A review of the British Post Office microwave radio-relay network. Pt. 3', *UK PO Elec. Eng. J.*, 1977, **70**, pp. 45–54

155 HARDEN, B. N., and TURNER, D.: 'Propagation studies and the development of terrestrial microwave radio-relay systems above 10 GHz in the United Kingdom' *in* 'Propagation of radio waves at frequencies above 10 GHz'. *IEE Conf. Publ. 98*, 1973, pp. 1–5

156 HEWITT, M. T., and NORBURY, J. R.: 'Correlation of fading on spaced microwave paths at 22 and 37 GHz' *in* 'Propagation of radio waves at frequencies above 10 GHz'. *IEE Conf. Publ. 98*, 1973, pp. 250–255

157 LEFRANCOIS, E., MARTIN, L., and ROORYCK, M.: 'Influence de la propagation sur le valeur de decouplage de deux polarisations orthogonales (Influence of propagation on the value of decoupling of two orthogonal polarizations)', *Ann. Telecommun.*, 1973

158 BATTESTI, J., and CONSTANTINIDIS, T.: 'Etude du découplage de polarisation sur une liaison de 53 km à 13 GHz: généralisation des résultats obtenus (Study of polarization decoupling on a path of 53 km at 13 GHz; generalisation of results obtained)'. Note Technique CNET-EST-APH-28

159 MORITA, K.: 'Fluctuations of cross polarisation discrimination ratio due to fading', *Rev. Electr. Commun. Lab.*, 1971, **19**

160 TURNER, D. J. W.: 'Measurements of cross polar discrimination at 22 and 37 GHz' *in* 'Propagation of radio waves at frequencies above 10 GHz'. *IEE Conf. Publ. 98*, 1973

161 SHIMBA, M., and MORITA, K.: 'Radio propagation characteristics due to rainfall at 19 GHz'. Proceedings of the International IEEE/G-AP Symposium, 1972, pp. 246–249

162 YAMAMOTO, H., MORITA, K., and NAKAMURA, Y.: 'Experimental considerations on 20 GHz high-speed digital radio relay systems' Proceedings of IEEE International Conference on Communications, 28-37-42, 1973

163 BARNETT, W. T.: 'Some experimental results on 18 GHz propagation', *ibid.*, 10E-1-4, 1972

164 NOWLAND, W. L., OLSEN, R. L., and SHKAROFSKY, I. P., 'Theoretical relationship between rain depolarization and attenuation', *Electron. Lett.*, 1977, **13**, pp. 676–678

165 HOSOYA, Y., and HASHIMOTO, A.: 'Rain induced depolarization of circularly polarized waves', *Rev. Electr. Commun. Labs.*, 1978, **26**

166 SEMPLAK, R. A.: 'The effect of rain on circular polarization at 18 GHz', *Bell Syst. Tech. J.*, 1973, **52**, pp. 1029–1031

167 DILWORTH, I. J., and EVANS, B. G.: 'Preliminary results of linear/circular depolarization on an 18 km, 11.6 GHz radio link', *Electron. Lett.*, 1976, **12**, pp. 618–620

168 YAMADA, M., OGAWA, A., FURUTA, O., and YUKI, H.: 'Rain depolarization measurement by using INTELSAT – IV satellite in 4-GHz band at a low elevation angle', *Ann. Telecommun.*, 1977, **32**, pp. 524–529

169 SCHULKIN, M.: 'Average radio-ray refraction in the lower atmosphere', *Proc. IRE*, 1952, **40**, pp. 554–561

170 CRANE, R. K.: 'Refraction effects in the neutral atmosphere'. *In* MEEKS, M.C. (Ed.) *Methods of experimental physics, Vol. 12 Astrophysics, Pt. B Radio Telescopes* (Academic Press, NY, 1976)

171 CRANE, R. K.: 'Propagation phenomena affecting satellite communication systems operating in the centimetre and millimetre wavelength bands', *Proc. IEEE*, 1971, **59**, pp. 173–188

172 MILLMAN, G. H.: 'A survey of tropospheric, ionospheric and extra-terrestrial effects on radio propagation between the Earth and space vehicles'. AGARD Conference Proceedings No. 3 (Propagation factors in space communication), pp. 3–55, Techivision, 1967

173 BRAMLEY, E. N.: 'Fluctuations in direction and amplitude of 136 MHz signals from a geostationary satellite', *J. Atmos. & Terr. Phys.*, 1974, **36**, pp. 1503–1513

174 GIBBINS, C. J., GORDON-SMITH, A. C., and CROOM, D. L.: 'Atmospheric emission measurements at 85 to 118 GHz', *Planet Space Sci.*, 1975, **23**, pp. 61–73

175 EMERY, R. J., MOFFAT, P., BOHLANDER, R. A., and GEBBIE, H. A.: 'Measurements of anomalous atmospheric absorption in the wavenumber range 4 cm^{-1}-15 cm^{-1}', *J. Atmos. & Terr. Phys.*, 1975, **37**, pp. 587–594

176 STRICKLAND, J. I.: 'The measurement of slant-path attenuation using radar, radiometers and a satellite beacon', *J. Rech. Atmos.*, 1974, **VIII**, pp. 347–358

177 BELL, R. R.: 'The calibration of 20 and 30 GHz radiometers using the ATS:6 satellite beacons'. Proceedings of the URSI Symposium on propagation in non-ionised media, La Baule, France, 1977, pp. 351–356

178 BRUSSAARD, G.: 'Rain attenuation on satellite-Earth paths at 11.4 and 14 GHz', *Ann. Telecommun.*, 1977, **32**, pp. 514–518

179 DAVIES, P. G.: 'Slant path attenuation at frequencies above 10 GHz'. *IEE Conf. Publ. 98*, April 1973, pp. 141–149

180 DAVIES, P. G.: 'Diversity measurements of attenuation at 37 GHz with sun-tracking radiometers in a 3-site network', *Proc. IEE*, 1976, **123**, pp. 765–769

181 YOKOI, H., YAMADA, M., and SATOH, T.: 'Atmospheric attenuation and scintillation of microwaves from outer space', *Publ. Astronom. Soc.*, 1970, **22**, pp. 511–524

182 YAMADA, M., and YOKOI, H.: 'Measurements of Earth-space propagation characteristics at 15.5 and 31.6 GHz using celestial radio sources', *Electron. & Commun.* 1974, **57-B**

183 CRANE, R. K.: 'Low elevation angle measurement limitations imposed by the troposphere: An analysis of scintillation observations made at Haystack and Millstone'. MIT Lincoln Lab. Tech. Dept. 518, Lexington, Mass., USA

184 STRICKLAND, J. I., OLSEN, R. L., and WERSTIUK, H. L.: 'Measurements of low angle fading in the Canadian arctic', *Ann. Telecommun.*, 1977, **32**, pp. 530–535

185 DAVIES, K.: *Ionospheric radio propagation* (NBS Monograph No. 80, 1965, also available from Dover Press)

186 COATES, R. J., and GOLDEN, T. S.: 'Ionospheric effects on telemetry and tracking signals from orbiting space-craft'. NASA TM-TX 63152, X-520-68-76 (N68-20043), Goddard Space Flight Center, Washington, DC, 1968

187 CHRISTIANSEN, R. M.: 'Preliminary report of S-band propagation disturbance during ASLEP mission support'. NASA X-861-71-239, Goddard Space Flight Center, Washington, DC, 1971

188 WHITNEY, H. E., AARONS, J., ALLEN, R. S., and SEEMANN, D. R.: 'Estimation of the cumulative amplitude probability distribution function of ionospheric scintillation', *Radio. Sci.*, 1972, **7**, pp. 1095–1104

189 FREMOUW, E. J., and RIND, C. L.: 'An empirical model for average F-layer scintillation at VHF/UHF', *ibid.*, 1973, **8**, pp. 213–220

190 FUNAKAWA, K., and OTSU, Y.: 'Characteristics of slant path rain attenuation at 35 GHz obtained by solar radiation and atmospheric emission
observations', *J. Rech. Atmos.*, 1974, 8

191 MORITA, K., and HIGUTI, I.: 'Statistical studies on rain attenuation and
site diversity effect on Earth-to-satellite links in microwave and millimetre
bands', *J. Inst. Electron. & Commun. Eng.* 1978, E-61

192 ALLNUTT, J. E., and SHUTIE, P. F.: 'Space diversity results at 30 GHz'.
Proceedings of the URSI Symposium on propagation in non-ionised media,
La Baule, France, 1977

193 WILSON, R. W., and MAMMEL, W. L.: 'Results from a three-radiometer
path-diversity experiment' *in* 'Propagation of radio waves at frequencies
above 10 GHz'. *IEE Conf. Publ. 98*, 1973

194 ALLNUTT, J. E.: 'Slant path attenuation and space diversity results using
11.6 GHz radiometers', *Proc. IEE*, 1976, **123**, pp. 1197–1200

195 ALLNUTT, J. E.: 'Variation of attenuation and space diversity with elevation angle on 12 GHz satellite-to-ground radio paths', *Electron. Lett.*,
1977, **13**, pp. 346–347

196 VOGEL, W. J., STRAITON, A. W., FANNIN, B. M., and WAGNER, N. K.:
'Attenuation diversity measurements at 20 and 30 GHz', *Radio Sci.*, 1976,
11, pp. 167–174

197 GRAY, D. A.: 'Earth-space path-diversity: dependence on base-line orientation'. Proceedings of the IEEE G-AP symposium, University of Colorado
Boulder, USA, pp. 366–369, 1973

198 HODGE, D. B.: 'Radar studies of rain attenuation and diversity gain'.
McGill University Stormy Weather Group Scientific Report MW-87, 1976

199 GOLDHIRSH, J.: 'Path attenuation statistics influenced by orientation of
rain cells', *IEEE Trans.*, 1976, **AP-24**, pp. 792–799

200 CRANE, R. K.: 'Morphology of ionospheric scintillation'. Technical note,
1974–29, Lincoln Lab. MIT, Lexington, Mass., USA, 1974

201 NOWLAND, W. L., STRICKLAND, J. I., SCHLESAK, J., and OLSEN,
R. L.: 'Measurements of depolarization and attenuation at 11.7 GHz using
the Communications Technology Satellite', *Electron. Lett.*, 1977, **13**, pp.
750–751

202 EVANS, B. G., and HOLT, A. R.: 'Scattering amplitudes and cross-polarisation of ice particles', *ibid.*, 1977, **13**, pp. 342–344

203 WATSON, P. A., McEWAN, N. J., DISSANAYAKE, A. W., HOWARTH,
D. P., and VAKILI, V. T.: 'Attenuation and cross-polarization measurements
at 20 GHz using the ATS-6 satellite with simultaneous radar observations'.
Proceedings of the URSI Symposium on Propagation in Non-Ionised Media,
La Baule, France, 1977

204 HOWELL, R. G.: 'Cross-polar phase variation at 20 GHz and 30 GHz on a
satellite-Earth path', *Electron. Lett.*, 1977, **13**, p. 405

205 McCORMICK, G. C., and HENDRY, A.: 'Depolarization by solid hydrometeors', *ibid.*, 1977, **13**, pp. 83–94

206 FIMBEL, J., and RAMAT, P.: 'Mesures de l'affaiblissement et de la depolarisation à 20 GHz a partir du satellite ATS 6 (Attenuation and depolarization measurement at 20 GHz from the ATS-6 satellite)', *Ann. Telecommun.*, 1977, **32**, pp. 497–501

207 NUSPL, P. P., DAVIES, N. G., and OLSEN, R. L.: 'Ranging and synchronisation accuracies in a regional TDMA experiment'. Proceedings of the
third International digital satellite communications conference, 1975

208 CRANE, R. K.: 'Coherent pulse-transmission through rain', *Trans. IEEE,* 1967, **AP-15,** pp. 252–256

209 National Bureau of Standards Technical Note No. 101, Revised, I and II, AD 687820 and AD 687821, NTIS, Springfield, Va., USA, 1967

210 HIRAI, M., KURIHARA, Y., INOUE, R., NIWA, S., IKEDA, M., and KIDO, Y.: 'Transmission loss in VHF and UHF overland propagation beyond the horizon', *J. Radio Res. Labs.,* 1963, **10**

211 HALL, M. P. M.: 'Statistics of high-level beyond-horizon signals at 2.2 GHz and 2.6 GHz, and measurements of the variation of the arrival-angle structure'. *In* 'Propagation effects on frequency sharing'. AGARD Conference Proceedings No. 127, May, 1973

212 EKLUND, F., and WICKERTS, S.: 'Wavelength dependence of microwave propagation far beyond the radio horizon', *Radio. Sci.,* 1968, **3,** p. 11

213 BOITHIAS, L., and BATTESTI, J.: 'Les faisceaux hertziens transhorizon de haute qualité (HIgh quality transhorizon radio-relay systems), *Ann. Telecommun.,* 1965

214 FEHLHABER, L., and GROSSKOPF, J.: 'Die mittlere schwundfrequenz auf scatterstrechen im frequenzbereich 1 GHz bis 10 GHz (The mean fading frequency of scatter links in the frequency range 1 GHz to 10 GHz)', *Tech. Ber. FTZ,* 1966, No. 5582

215 ABEL, N.: 'Beobachtungen an einer 210 km langen, 12-GHz-Streche (Observations on a 12 GHz link of 210 km length)', *Tech. Ber. FTZ,* 1972, No. A455, TBr34

216 WRIGHT, K. F., and COLE, J. E.: 'Measured distribution of the duration of fades in tropospheric scatter transmission', *Trans. IRE,* 1960, **AP-8,** pp. 594–598

217 VIGANTS, A.: 'Number and duration of fades at 6 and 4 GHz', *Bell. Syst. Tech. J.,* 1971, **50,** pp. 815–841

218 COX, D. C., and WATERMAN, A. T. JUR.: 'Phase and amplitude measurements of transhorizon microwaves with a multi-data-gathering antenna array'. *In* 'Scatter propagation of radio waves'. AGARD Conference Proceedings No. 37, August, 1968

219 GJESSING, D. T., JESKE, H., and KLINT HANSEN, N.: 'An investigation of the tropospheric fine scale properties using radio, radar and direct methods', *J. Atmos. & Terr. Phys.,* 1969, **31,** pp. 1157–1182

220 LAMMERS, U. H. W., and OLSEN, R. L.: 'Bistatic measurement of meteorological and propagation parameters with high resolution K_u – band scatter system'. IEEE G-AP International Symposium, Boulder, Colo, USA, August 1973 (New York, USA: IEEE 1973), pp. 200–203

221 STARAS, H.: 'Antenna-to-medium coupling loss', *Trans. IRE,* 1957, **AP-5,** pp. 228–231

222 GOUGH, M. W.: 'Aperture-medium coupling loss in troposcatter propagation – an engineering appraisal', *in* 'Tropospheric wave propagation'. *IEE Conf. Publ. 48,* 1968, pp. 93–100

223 HALL, M. P. M., and MISME, P.: 'Gain degradation of a 25-m parabaloidal aerial on 2 GHz transhorizon radio paths', *Proc. IEE,* 1975, **122,** pp. 358–360

224 BOITHIAS, L., and BATTESTI, J.: 'Nouvelles experimentations sur la baisse de gain d'antenne dans les liaisons transhorizon, (New experiments on loss of antenna gain in transhorizon links)' *Ann. Telecommun.,* 1967, **22,** pp. 321–325

225 BATTESTI, J., and BOITHIAS, L.: 'Nouveaux éléments sur la propagation par les hétérogénéites de l'atmosphère (Contribution to the study of propagation by atmospheric discontinuties)', *ibid.*, 1971

226 GERKS, I. H.: 'Factors affecting spacing of radio terminals in a uhf link', *Proc. IRE*, 1955, **43**, pp. 1290–1297

227 SHAFT, P. D.: 'Information bandwidth of tropospheric scatter systems', *IRE Trans.*, 1961, **CS-9**, pp. 280–287

228 IWASA, H., and KUROBE, A.: 'Experimental results of angle diversity effect in trans-horizon radio relay system', *Trans. IECE Japan*, 1974, **57-B**, pp. 592–593

229 GOUGH, M. W., and RIDER, G. C.: 'Angle diversity in troposcatter communications: some confirmatory trials', *Proc. IEE*, 1975, **122**, pp. 713–719

230 SURENIAN, D.: 'Experimental results of angle diversity system tests', *IEEE Trans.*, 1965, **COM-13**, p. 208

231 DOUGHERTY, H. T., and WILKERSON, R. E.: 'Determination of antenna height for protection against microwave diffraction fading', *Radio Sci.*, 1967, **2**, pp. 161–165

232 WAIT, J. R., and CONDA, A. M.: 'Diffraction of electromagnetic waves by smooth obstacles for grazing angles', *J. Res. N. B. S.*, 1959, **63D**, pp. 181–197

233 DOUGHERTY, H. T., and MALONEY, L. J.: 'Application of diffractions by convex surfaces to irregular terrain situations', *Radio Sci.*, 1964, **68D**, pp. 239–250

234 MILLINGTON, G., HEWITT, R., and IMMIRZI, F. S.: 'Double knife-edge diffraction in field strength predictions', *IEE Monograph 507E*, 1962, **109**, Part C, p. 419

235 DEYGOUT, J.: 'Multiple knife-edge diffraction of microwaves', *IEEE Trans.*, 1966, **AP-14**, pp. 480–489

236 ASSIS, M. S.: 'A simplified solution to the problem of multiple diffraction over rounded obstacles', *ibid.*, 1971, **AP-19**, pp. 292–295

237 FURUTSU, K.: 'Wave propagation over an irregular terrain. Parts I and II', *J. Rad. Res. Labs. Japan*, 1957, **4**, pp. 16 and 18

238 FURUTSU, K.: 'Wave propagation over an irregular terrain. Part III', *ibid.*, 1959, **6**, p. 23

239 HUFFORD, G. A.: 'An integral equation approach to the problem of wave propagation over an irregular terrain', *Q. J. Appl. Maths.*, 1952, **9**, pp. 391–404

240 DICKSON, F. H., EGLI, J. J., HERBSTREIT, J. W., and WICKIZER, G. S.: 'Large reductions of VHF transmission loss and fading by the presence of a mountain obstacle in beyond-line-of-sight paths', *Proc. IRE*, 1953, pp. 967–969

241 NORTON, K. A., *et al.*: 'Use of angular distance in estimating transmission loss and fading range for propagation through a turbulent atmosphere over irregular terrain', *ibid.*, 1955, **43**, p. 1488

242 KING, R. W., and PAGE, H.: 'The propagation of electromagnetic waves over irregular terrain' *in* Propagation effects on frequency sharing'. *AGARD Conference Proceedings 127*, 1973

243 FURUTSU, K.: 'Effect of ridge, cliff and bluff at a coastline on ground-waves', *J. Radio Res. Labs. Japan*, 1962, **9**, p. 41

244 FURUTSU, K., and WILKERSON, R. E.: 'Optical approximation for the residue series of terminal gain in radio wave propagation over inhomogeneous earth', *Proc. IEE*, 1971, **118**, pp. 1197–1202

245 IWAI, F., *et al.:* 'Radio transmission beyond the line-of-sight, etc.' *Electr. Commun. Lab. Tech. J.*, 1959, **8**, p. 1183

246 FUKAMI, T., *et al.:* 'Propagation measurements on long mountain diffraction paths, etc.', *ibid.*, 1961, **10**, p. 2430

247 KIRBY, R. S., DOUGHERTY, H. T., and McQUATE, P. L.: 'Obstacle gain measurements over Pikes Peak at 60 to 1046 Mc', *Proc. IRE*, 1955, **43**, pp. 1467–1472

248 CAUSEBROOK, J. H.: 'Computer prediction of UHF broadcast service areas', BBC Research Deptartment Report No. RD 1974/4

249 DURKIN, J.: 'Computer prediction of service areas for VHF and UHF land mobile radio services', *IEEE Trans.*, 1977, **VT-26**, pp. 323–327

250 LONGLEY, A. G., and RICE, P. L.: 'Prediction of tropospheric radio transmission loss over irregular terrain, a computer method – 1968'. ESSA Technical Report ERL 79-ITS 67, Access No. 676 874, NTIS, Springfield Va, USA

251 OKUMURA, Y., *et al.:* 'Field strength and its variability in VHF and UHF land-mobile radio services', *Rev. Electr. Commun. Lab.*, 1968, **16**, pp. 825–873

252 KINASE, A.: 'Influence of terrain irregularities and environmental clutter surroundings on the propagation of broadcasting waves in the UHF and VHF bands' NHK (Japan Broadcasting Corporation), Technical Memo 14, 1969, pp. 1–64

253 LONGLEY, A. G.: 'Radio propagation in urban area'. OT Report 78-144, NTIS, Springfield, Va, USA, April 1978

254 LONGLEY, A. G.: 'Location variability of transmission loss for land mobile and broadcast systems'. OT Report 76-87 NTIS, Springfield, Va, USA, May 1976

255 SOFAER, E., and BELL, C. P.: 'Factors affecting the propagation and reception of broadcasting signals in the UHF bands', *Proc. IEE*, 1966, **113**, p.7

256 REUDINK, D. O., and BLACK, D. M.: 'Some characteristics of mobile radio propagation at 836 MHz in the Philadelphia area', *IEEE Trans.*, 1972, **VT-21**

257 JAKES, W. C.: *Microwave mobile communications* (Wiley, 1974)

258 KOSONO, S., and WATANABE, K.: 'Influence of environmental buildings on land mobile radio propagation', *IEE Trans.*, 1964 **COM-10**, p. 6

259 BURROWS, C. R., HUNT, L. E., and DECINO, A.: 'Ultra-short-wave propagation: Mobile urban transmission characteristics', *Bell Syst. Tech. J.*, 1935, **14**, pp. 253–272

260 RICE, P. L.: 'Radio transmission into buildings at 35 and 150 MHz', *ibid.*, 1959, **38**, pp. 197–210

261 DURANTE, J. M.: 'Building penetration loss at 900 MHz'. *IEEE Vehicular TEchnology Conference Record*, 1973

262 SHAFER, J.: 'Propagation statistics of 900 MHz and 450 MHz signals inside buildings'. Proceedings of the Microwave Mobile Radio Symposium, Boulder, Colorado, March 1973

263 BELL, C. P.: 'UHF field strength distributions at typical domestic receiving locations', BBC Research Department Report K-168, 1964/2

264 JONES, L. F.: 'A study of the propagation of wavelengths between three and eight metres', *Proc. IRE,* 1933, **21,** pp. 349–386

265 BARTON, F. A., and WAGNER, G. A.: 'What happens when 900 MHz and 450 MHz takes to the hills', *Communications,* 1974 March, pp. 20–24 and April, pp. 22–25

266 WAIT, J. R. *et al.* (Eds.): 'Workshop on radio systems in forrested and/or vegetated environments'. USACC Technical Report No. ACC-ACO-1-74 (AD 780712), NTIS, Springfield, Va, USA, 1974

267 TAMIR, T.: 'Radio wave propagation along mixed paths in forest environments', *IEE Trans.,* 1977, **AP-25,** pp. 471–477

268 SACHS, D. L. *et al.:* 'A conducting slab model for electromagnetic propagation within a jungle medium', *Radio Sci. (New Series),* 1968, **3,** pp. 125–134

269 SAXTON, J. A., and LANE, J. A.: 'VHF and UHF reception. Effects of trees and other obstacles'. *Wireless World,* 1955, **61,** pp. 229–232

270 JOSEPHSON, B., and BLOMQUIST, A.: 'The influence of moisture in the ground, temperature and terrain on ground-wave propagation in the VHF band', *IRE Trans.,* 1958, **AP-6,** p. 169

271 HEAD, H. T.: 'The influence of trees on television field-strengths at UHF', *Proc. IRE,* 1960, **48,** p. 1016

272 HERBSTREIT, J. W., and CRICHLOW, W. Q.: 'Measurements of the attenuation of radio signals by jungles', *Radio Sci.,* 1964, **68D,** p. 903

273 BLOMQUIST, A.: 'Depolarisation of radio-waves in terrain'. Res. Instit. of National Defence, Stockholm, Report A648, 1965

274 EPSTEIN, J., and PETERSON, D. W.: 'An experimental study of wave propagation at 850 Mc', *Proc. IRE,* 1953, **41,** pp. 595–611

275 YOUNG, W. R.: 'Comparison of mobile radio transmission at 150, 450, 990 and 3700 Mc/s', *Bell Syst. Tech. J.,* 1952, **31,** pp. 1068–1085

276 PARKER, R. E. and ROPER, G. B.: 'Vehicular radio communication in London'. SDE/X/B70/1, AWRE Aldermaston, February 1970

277 NYLUND, H. W.: 'Characteristics of small-area signal fading on mobile circuits in the 150 MHz band', *IEEE Trans.,* 1968, **VT-17,** pp. 24–30

278 FINE, H.: 'Variation of field intensity over irregular terrain within line-of-sight for the UHF band', *ibid.,* 1952, **AP-4,** pp. 53–65

279 ALLSEBROOK, K., and PARSONS, J. D.: 'Mobile radio propagation in British cities at frequencies in the VHF and UHF bands', *ibid.,* 1977, **VT-26,** pp. 313–323

280 CLARKE, R. H.: 'A statistical theory of mobile radio reception', *Bell Syst. Tech. J.,* 1968, **47**

281 GANS, M. J.: 'A power-spectral theory of propagation in the mobile-radio environment', *Trans. IEEE,* 1972, **VT-21,** pp. 27–38

282 HENZE, M., and PARSONS, D.: 'Experimental dual-diversity single-receiver predetection combiner for u.h.f. mobile radio', *IEE J. Electron. Circuits & Syst.,* 1976, **1,** pp. 2–10

283 PARSON, J. D., HENZE, M., RATCLIFF, P. A., and WITHERS, M. J.: 'Diversity techniques for mobile radio reception', *IEEE Trans.,* 1976, **VT-25,** pp. 75–85

284 JAKES, W. C. J.: 'A comparison of specific space diversity techniques for reduction of fast fading in UHF mobile radio systems', *ibid.,* 1971, **VT-20,** pp. 81–92

285 YOUNG, W. R., and LACY, L. Y.: 'Echoes in transmission at 450 mega-cycles from land-to-car radio units', *Proc. IRE*, 1950, **38**, pp. 255–258

286 LEE, W. C. Y.: 'Preliminary investigation of mobile radio signal fading using directional antennae on the mobile unit', *IEEE Trans.*, 1966, **VT-15**, pp. 8–15

287 STIDHAM, J. R.: 'Experimental study of UHF mobile radio transmission using a directive antenna', *ibid.*, 1966, **VC-15**, pp. 16–24

288 HAGN, G. H.: 'Radio noise of terrestrial origin', *Radio Sci.*, 1973, **8**, pp. 613–621

289 HERMAN, J. R.: 'Survey of man-made radio noise', *Prog. Radio Sci.*, 1971, **1**, pp. 315–348

290 HORNER, F.: 'Techniques used for the measurement of atmospheric and man-made noise', *ibid.*, 1971, II, pp. 177–182

291 SPAULDING, A. D., and DISNEY, R. T.: 'Man-made noise, Pt I, OT Report 74-38, US Government Printing Office, Washington DC 20402, 1974

292 SPAULDING, A. D., DISNEY, R. T., and HUBBARD, A. G.: 'Man-made noise, Pt. II', OT Report 75–63, US Government Printing Office, Washington, DC 20402, 1975

293 KING, R. W., and CAUSEBROOK, J. H.: 'Computer programs for UHF co-channel interference prediction using a terrain computer bank'. BBC Research Department Report No. 1974/6

294 GRAY, R. E.: 'Refractive index of the atmosphere as a factor in tropo-spheric propagation far beyond the horizon', *Electr. Commun.*, 1959, **36**, p. 60

295 BOITHIAS, L., and MISME, P.: 'Limitations on the application of the re-fractive index at the surface of the earth', *Ann. Telecommun.*, 1964, **19**

296 BATTESTI, J., BOITHIAS, L., and MISME, P.: 'Calcul des affaiblissements en propagation transhorizon à partir des paramètres radiométéorologiques (Calculations of attenuation in transhorizon propagation from radio-meteor-ological parameters)', *ibid.*, 1968, May-June

297 TROITSKY, N. V.: 'Fading of ultra-short waves in radio-relay systems', *Elektrosviaz*, 1958, **10**

298 CRANE, R. K.: 'Bistatic scatter from rain', *IEEE Trans.*, 1974, **AP-22**, pp. 312–320

299 LARI, G., TAGHOLM, L. F., and BELL, C. P.: 'Research on VHF iono-spheric propagation (band 1)'. Techn. 3085-E, EBW Technical Centre, Brussels, 1967

References 301 to 325 are Recommendations and Reports taken from the CCIR volumes published by the International Telecommunications Union, 2 Rue Varembe 1211, Geneva 20, Switzerland. The volume numbers are as follows:

I	Spectrum utilisation and monotoring	(Study Group 1)
IV	Fixed service using communication satellites	(Study Group 4)
V	Propagation in non-ionised media	(Study Group 5)
VI	Ionospheric propagation	(Study Group 6)
IX	Fixed service using radio-relay systems	(Study Group 9)
XI	Broadcasting service (television) including video-recording and satellite applications	(Study Group 11)

301 'The concept of transmission loss in studies of radio systems', Recommen-dation 341, **Vol. I**

302 'VHF and UHF propagation curves for the frequency range from 30 MHz to 1000 MHz. *Broadcasting services'*, Recommendation 370, **Vol. V**

303 'Electrical characteristics of the surface of the Earth', Recommendation 527, **Vol. V**

304 'VHF, UHF and SHF propagation curves for the aeronautical mobile service', Recommendation 528, **Vol. V**

305 'Broadcasting satellite service: sound and television', Report 215, **Vol. XI**

306 'Measurement of field strength for VHF (metric) and VHF (decimetric) broadcast services, including television', Report 228, **Vol. V**

307 'Influence of terrain irregularities and vegetation on tropospheric propagation', Report 236, **Vol. V**

308 'Propagation data required for transhorizon radio-relay systems', Report 238, **Vol. V**

309 'Propagation statistics required for braodcasting services, using the frequency range 30 to 1000 MHz', Report 239, **Vol. V**

310 'Man-made radio noise', Report 258, **Vol. VI**

311 'VHF propagation by regular layers, sporadic-E or other anomalous ionisation', Report 259, **Vol. VI**

312 'Ionospheric effects upon Earth-space propagation', Report 263, **Vol. VI**

313 'Trans-horizon radio-relay systems', Report 285, **Vol. IX**

314 'Propagation data required for line-of-sight radio-relay systems', Report 338, **Vol. V**

315 'Polarization discrimination by means of orthogonal circular and linear polarization', Report 555, **Vol IV**

316 'Radiometeorological data', Report 563, **Vol. V**

317 'Propagation data required for space telecommunication systems', Report 564, **Vol. V**

318 'Propagation curves and statistics required for land mobile services using the frequency range 30 MHz to 1 GHz', Report 567, **Vol. V**

319 'The evaluation of propagation factors in interference problems at frequencies greater than about 0.6 GHz', Report 569, **Vol. V**

320 'Propagation by diffraction', Report 715, **Vol. V**

321 'Effects of tropospheric refraction on radio wave propagation', Report 718, **Vol. V**

322 'Attenuation by gases', Report 719, **Vol. V**

323 'Radio emission due to absorption by atmospheric gases and precipitation', Report 720, **Vol. V**

324 'Attenuation and scattering by rain and other atmospheric particles', Report 721, **Vol. V**

325 'Propagation data for the evaluation of co-ordination distance in the frequency range 1-40 GHz', Report 724, **Vol. V**

Index

Many topics are indexed under the specific path types: earth-space paths, line-of-sight paths and transhorizon paths